Studies in Systems, Decision and Control

Volume 693

The series "Studies in Systems, Decision and Control" (SSDC) covers both new developments and advances, as well as the state of the art, in the various areas of broadly perceived systems, decision making and control–quickly, up to date and with a high quality. The intent is to cover the theory, applications, and perspectives on the state of the art and future developments relevant to systems, decision making, control, complex processes and related areas, as embedded in the fields of engineering, computer science, physics, economics, social and life sciences, as well as the paradigms and methodologies behind them. The series contains monographs, textbooks, lecture notes and edited volumes in systems, decision making and control spanning the areas of Cyber-Physical Systems, Autonomous Systems, Sensor Networks, Control Systems, Energy Systems, Automotive Systems, Biological Systems, Vehicular Networking and Connected Vehicles, Aerospace Systems, Automation, Manufacturing, Smart Grids, Nonlinear Systems, Power Systems, Robotics, Social Systems, Economic Systems and other. Of particular value to both the contributors and the readership are the short publication timeframe and the world-wide distribution and exposure which enable both a wide and rapid dissemination of research output.

Indexed by SCOPUS, DBLP, WTI Frankfurt eG, zbMATH, SCImago.

All books published in the series are submitted for consideration in Web of Science.

Shyam Kamal • Sunidhi Pandey

A Continuous-Time Perspective on Accelerated Optimization and Control

Springer

Shyam Kamal
Electrical Engineering
IIT (BHU) Varanasi
Varanasi, Uttar Pradesh, India

Sunidhi Pandey
Department of Electrical Engineering
IIT (Banaras Hindu University)
Varanasi, Uttar Pradesh, India

ISSN 2198-4182 ISSN 2198-4190 (electronic)
Studies in Systems, Decision and Control
ISBN 978-3-032-31644-8 ISBN 978-3-032-31645-5 (eBook)
https://doi.org/10.1007/978-3-032-31645-5

This Springer imprint is published by the registered company Springer Nature Switzerland AG
The registered company address is: Gewerbestrasse 11, 6330 Cham, Switzerland

To our family (Aprajita, Adhita, Avyaan, Sushila, Pratibha Pandey, Vijay Kumar Pandey, Prashant Pandey and Vibhav Pandey), whose unwavering patience, understanding, and steadfast encouragement formed the foundation of this work. Through the many challenges and sustained efforts along this journey, your constant belief in us remained a source of strength and clarity. This accomplishment stands as much a reflection of your support as it does of our dedication to the pursuit of knowledge.

Preface

Optimization, in its most general form, concerns the systematic determination of decision variables that extremize a prescribed performance functional subject to given constraints. Classical approaches typically formulate such problems in finite-dimensional spaces, aiming to determine an extremizer of a scalar-valued objective over an admissible set [1, 2]. These approaches typically emphasize algorithmic procedures and convergence analysis in discrete-time settings.

In contrast, an alternative and increasingly influential paradigm interprets optimization through the lens of continuous-time (CT) dynamical systems. Early connections between optimization and differential equations can be traced to works such as Variational Analysis [3, 4] and Optimization by Vector Space Methods [5]. More recent developments have formalized optimization algorithms as CT flows, notably through gradient flow (GF) interpretations and related dynamical system frameworks [6]. From this perspective, the optimizer is characterized as an equilibrium (or invariant set) of an associated dynamical system, and the optimization process is understood as the asymptotic evolution of system trajectories governed by differential equations [6, 7].

A central construct in continuous-time optimization (CTO) is the representation of the decision process as a dynamical system governing the temporal evolution of the decision variables. Formally, the dynamics are described by $\dot{x}(t) = F(x(t)), \quad x(t) \in \mathbb{R}^n$, where the vector field F is systematically designed such that its equilibrium points concur with the solution set of the underlying optimization problem. This establishes a direct correspondence between optimality conditions and stability properties of the induced dynamical system [6, 8]. Equally fundamental is the role of the objective function, which encodes the performance criterion to be minimized. Unlike in classical static formulations—where the objective merely serves as a scalar measure of optimality—here it assumes an intrinsic and generative role in shaping the system dynamics. In particular, the vector field F is often constructed explicitly from the structure of the objective function, for instance via GF or, more generally, through operator-theoretic formulations involving monotone operators [9, 10]. This close interplay ensures that the system evolution is inherently aligned with descent directions, thereby driving trajectories toward optimality in a principled and structured manner.

First-order methods play a central role due to their scalability and computational efficiency in large-scale settings [11]. In particular, gradient-based algorithms are widely adopted for their robustness to noise, ease of implementation, and low memory requirements [12]. Methods for locating minima (local) of a function $g(x)$ over an unconstrained domain $x \in \mathbb{R}^n$ date back to the origins of differential calculus and are fundamentally based on following descent directions. The most canonical class comprises gradient (or steepest descent) methods, exemplified by the dynamical system $\dot{x} = -\nabla g(x)$. These methods admit numerous variants and extensions, and are not restricted to Euclidean spaces; they naturally generalize to non-Euclidean settings, including optimization on manifolds [6, 7, 13, 14]. Continuous-time negative GF, convergence to minimizers appears plausible, if not tautological [15]. A natural degree of freedom in GF formulations arises from the introduction of a state-dependent metric $P(\cdot)$, which assigns to each point $x \in \mathbb{R}^n$ a symmetric positive-definite (PD) matrix $P(x)$. Under mild regularity assumptions—such as a uniformly bounded condition number of $P(x)$ [16]—the trajectories of the generalized GF $\dot{x} = -P(x)\nabla g(x)$ converge to the set of critical points of g. In particular, the incorporation of the metric $P(x)$ preserves the equilibrium set of the system, i.e., it does not alter the location of the critical points. As a particular instance of a state-dependent metric, suppose that g is twice continuously differentiable and strongly convex. In this case, the metric $P(x)$ may be chosen as the inverse of the Hessian, i.e., $P(x) = \nabla^2 g(x)^{-1}$. This choice gives rise to the CT analogue of the classical Newton method, commonly referred to as the Newton flow (NF) [17].

However, in contrast to its discrete-time counterpart, the CT-NF does not inherently yield an accelerated convergence rate relative to standard GF. Instead, the use of the inverse Hessian induces isotropic convergence behavior, meaning that the rate of convergence is uniform across all directions. This property effectively compensates for the ill-conditioning of the objective function.

In a broad class of engineering applications—including robotics, power networks, and autonomous systems—decision-making must be performed in real time, thereby rendering convergence within a prescribed temporal horizon a critical design requirement. This has motivated a shift beyond asymptotic notions of optimality toward stronger, time-constrained convergence guarantees. To this end, recent advances have focused on optimization and control algorithms endowed with finite-time, fixed-time, and predefined-time stability (PTS) properties. Finite-time stability (FTS) ensures convergence within a finite interval that typically depends on the initial conditions [18–21], whereas fixed-time stability (FxTS) ensures convergence within a uniform time bound that does not depend on the initial state [22, 23]. Extending these notions, PTS enables the explicit assignment of the convergence time a priori, thereby providing precise control over the transient behavior of the system [24–28]. Such guarantees are particularly well-suited for time-critical applications, where adherence to stringent temporal specifications is as essential as achieving optimal performance.

A central theme of this monograph is the systematic integration of optimization with control-theoretic principles. By leveraging tools such as Lyapunov functions, comparison lemmas, and input-to-state stability (ISS) [29, 30], CTO dynamics can be designed to ensure not only convergence but also robustness against disturbances and uncertainties. Beyond asymptotic and finite-time convergence (FTC) [31], particular emphasis is placed on predefined-time convergence (PTC), wherein the settling time can be explicitly prescribed a priori, independent of initial conditions [32]. This feature is especially significant in real-time and safety-critical applications, as it enables guaranteed performance within a user-defined time horizon while preserving robustness properties of the closed-loop optimization dynamics.

Most real-time optimization problems encountered in practical applications are inherently constrained, arising from physical limitations, safety requirements, or operational bounds. Consequently, it becomes essential to develop systematic frameworks that can effectively handle such constraints while preserving desirable convergence properties.

Within the CTO paradigm, constrained problems can be rigorously addressed by embedding the feasible set geometry into the dynamical system through projection operators or related mechanisms. In particular, when the constraint set admits efficient projection, projected gradient flows (PGF) provide a natural extension of unconstrained dynamics, ensuring that trajectories remain within the feasible region at all times. In this context, consider the constrained optimization problem $\min_{x \in \mathcal{X}} \ g(x)$, where $\mathcal{X} \subset \mathbb{R}^n$ is a non-empty, closed, and convex set, and $g : \mathbb{R}^n \to \mathbb{R}$ is a continuously differentiable function. In a broad sense, under mild regularity conditions, solutions of constrained optimization problems can be characterized as saddle points of an associated Lagrangian function. However, classical GF dynamics, as introduced earlier, are not inherently equipped to handle unilateral constraints. A common and practically effective approach to address this limitation is to augment the objective function with suitably designed penalty or barrier terms, thereby transforming the constrained problem into an equivalent unconstrained formulation while implicitly enforcing feasibility.

In this monograph, constrained optimization problems are addressed within the CT framework by systematically integrating projection-based and barrier-based methodologies. In particular, equality constraints are handled using projected dynamical systems, ensuring that the system trajectories evolve within the feasible manifold and optimality is achieved within a user-specified time [33].

An important and closely connected theme in contemporary optimization is the analysis of time-varying or non-stationary problems, which are central to online and real-time decision processes. More recently, this area has attracted considerable attention due to its wide range of applications in control systems, robotics, and machine learning [34–36]. The primary objective in such problems is not merely to compute an optimal solution, but to design algorithms capable of accurately tracking a time-evolving optimizer while providing rigorous performance guarantees.

Broadly, two complementary perspectives have emerged in the literature. The first views time-varying optimization (TVO) as an online learning problem, where decisions are updated iteratively based on revealed objective functions, with the goal of minimizing cumulative regret [37, 38]. The second, more aligned with control-theoretic principles, interprets the problem as a trajectory tracking task, wherein the optimizer defines a time-dependent solution manifold that must be closely followed by the optimization dynamics [39, 40]. Despite their differing viewpoints, both approaches rely on a common set of structural assumptions to ensure well-posedness and tractability. In particular, convexity (and often strong convexity) is typically imposed to ensure the existence and uniqueness of the optimizer, while bounded variation assumptions on the objective and constraint functions are introduced to enable meaningful tracking guarantees. Depending on the application, such TVO problems may be unconstrained, constrained over static feasible sets, or involve explicitly time-dependent constraints, each introducing additional layers of complexity.

Within this monograph, TVO is primarily approached from a dynamical systems and control-theoretic perspective, where the objective is to design CT algorithms that can track the optimal solution with PTC guarantees, even in the presence of uncertainties and time-varying constraints [41].

Traditionally, optimization has been treated as a numerical and algorithmic discipline, wherein problems are solved offline, and the computed solutions are subsequently deployed as set-points for physical systems. While this paradigm has been widely successful, it inherently assumes a clear separation between optimization and system dynamics. In practice, however, such a separation can lead to degraded performance in the presence of modeling uncertainties, external disturbances, and time-varying environments, where static solutions quickly become suboptimal. To overcome these limitations, the paradigm of feedback-based optimization (FBO) [15, 42], also referred to as autonomous optimization, has emerged as a powerful alternative. In this framework, optimization dynamics are embedded directly within the feedback loop of the physical system, enabling real-time adaptation based on measurements. From this perspective, optimization algorithms can be interpreted as dynamical systems interacting with the plant, or equivalently, as feedback controllers that inherently steer the system toward optimal operating conditions. Early manifestations of this idea can be found in communication networks and power systems, where system dynamics were shown to implicitly solve underlying optimization problems.

A fundamental challenge in FBO lies in ensuring the stability of the resulting closed-loop interconnection between the plant and the optimization dynamics. Unlike conventional optimization algorithms, the bidirectional coupling between system and algorithm introduces additional complexity that may lead to instability or slow convergence. To address this, the concept of timescale separation has been extensively investigated, wherein stability is achieved by designing the optimization dynamics to evolve on a slower timescale relative to the plant, allowing the latter to be approximated by its steady-state response. This insight is rigorously supported by singular perturbation theory, which provides explicit conditions for stability of such interconnected systems. While related methods such as extremum seeking control [43] and real-time optimization [44] share similar objectives, they often rely on probing signals, numerical solvers, or model-based predictions. In contrast, FBO directly exploits system dynamics to achieve optimality without explicitly solving an optimization problem at each iteration. Despite these advances, most existing methods primarily guarantee asymptotic or exponential convergence, which may be insufficient for time-critical applications [42]. This limitation motivates the integration of finite-time, fixed-time, and particularly predefined-time convergence concepts within FBO frameworks, an area that remains relatively underexplored and forms an interesting section of this monograph [45].

Motivated by the growing demand for fast, reliable, and real-time decision-making in modern engineering systems, this monograph seeks to bridge the gap between classical optimization theory and control-oriented dynamical system approaches. While traditional optimization methods predominantly offer asymptotic convergence guarantees and operate in offline settings, many practical applications—ranging from robotics and power systems to autonomous and networked systems—require solutions that can be computed and implemented within strict time constraints, while remaining robust to uncertainties and disturbances. Furthermore, the increasing relevance of time-varying and constrained optimization problems, coupled with the emergence of FBO frameworks, necessitates the development of algorithms that can seamlessly integrate with system dynamics. In this context, the primary motivation of this monograph is to develop a unified CTO framework with strong convergence guarantees, particularly emphasizing predefined-time convergence (PTC), enabling explicit control over convergence time while ensuring robustness, feasibility, and real-time applicability. The subsequent chapters of the monograph present the core contributions, summarized as follows:

- Chapter 4 presents the development of CT dynamical systems for solving both unconstrained and equality-constrained optimization problems, with guaranteed convergence within a predefined time horizon under the assumption that the objective functions satisfy a gradient dominance condition. The convergence properties of the proposed dynamics are rigorously established using Lyapunov-based analysis. In addition, the robustness of these GF in the presence of perturbations is investigated through the framework of ISS. The effectiveness and practical applicability of the proposed methods are further illustrated by simulation studies on least-squares and resource allocation problems.
- Chapter 5 develops a class of second-order dynamical systems for addressing TVO problems, with the objective of tracking the optimal trajectory within a predefined time. The chapter begins by formulating a Newton-like dynamical system for trajectory tracking in constrained convex TVO problems. To accommodate scenarios where the time derivative of the gradient is unavailable or uncertain, a robust Newton-inspired dynamic ensuring PTS is discussed. Moreover, to tackle cases involving singular or non-invertible Hessians, we study a Levenberg—Marquardt-type dynamical system with PTC properties.
- Moreover, the sixth chapter investigates the practical applicability of the dynamical frameworks developed in Chap. 5. In particular, the studied methods are

employed for robot navigation with collision avoidance, where the objective is to track a moving target within a predefined time horizon. The problem is formulated as a TVO task with inequality constraints, ensuring convergence to the target from almost all initial conditions. Numerical simulations demonstrate the robustness and effectiveness of the proposed dynamics in complex and obstacle-dense environments.

- In the final chapter, our interest moved toward integrating feedback-based CTO with finite-time and fixed-time convergence (FxTC) properties. Optimization problems are formulated within a dynamical systems framework, wherein the decision variables evolve in closed-loop interaction with the underlying plant dynamics. A unified Lyapunov-based framework is developed, employing composite Lyapunov functions to capture the coupled behavior of accelerated gradient flow dynamics and system dynamics. The proposed theoretical results are substantiated through a nonlinear coupled-tank system, illustrating enhanced convergence speed and improved robustness in comparison with conventional approaches.

References

1. Boyd, S.P., Vandenberghe, L.: Convex Optimization. Cambridge University Press (2004)
2. Nocedal, J., Wright, S.J.: Numerical Optimization. Springer (2006)
3. Rockafellar, R.T., Wets, R.J.: Variational Analysis. Springer (1998)
4. Pandey, S., Vishwakarma, A., Kamal, S., Saket, R.: Variational methods in analytic number theory: A unified framework inspired by Ramanujan's work. In: 2025 Eleventh Indian Control Conference (ICC), pp. 727–732. IEEE (2025)
5. Luenberger, D.G.: Optimization by Vector Space Methods. John Wiley & Sons (1997)
6. Helmke, U., Moore, J.B.: Optimization and Dynamical Systems. Springer Science & Business Media (2012)
7. Absil, P.-A., Kurdyka, K.: On the stable equilibrium points of gradient systems. Syst. Control Lett. **55**(7), 573–577 (2006)
8. Khalil, H.K.: Nonlinear Systems, 3rd edn. Prentice-Hall, Upper Saddle River (2002)
9. Su, W., Boyd, S., Candes, E.J.: A differential equation for modeling nesterov's accelerated gradient method: Theory and insights. J. Mach. Learn. Res. **17**(153), 1–43 (2016)
10. Attouch, H., Peypouquet, J., Redont, P.: A dynamical approach to an inertial forward-backward algorithm for convex minimization. SIAM J. Optim. **24**(1), 232–256 (2014)
11. Nesterov, Y.: Introductory Lectures on Convex Optimization: A Basic Course, vol. 87. Springer Science & Business Media (2013)

12. Chen, F., Ren, W.: Sign projected gradient flow: A continuous-time approach to convex optimization with linear equality constraints. Automatica **120**, 109156 (2020)
13. Brockett, R.W.: Dynamical systems that sort lists, diagonalize matrices, and solve linear programming problems. Linear Algebra Appl. **146**, 79–91 (1991)
14. Zehnder, E.: Lectures on Dynamical Systems: Hamiltonian Vector Fields and Symplectic Capacities, vol. 11. European Mathematical Society (2010)
15. Hauswirth, A., He, Z., Bolognani, S., Hug, G., Dörfler, F.: Optimization algorithms as robust feedback controllers. Annu. Rev. Control **57**, 100941 (2024)
16. Hauswirth, A., Bolognani, S., Dorfler, F.: Projected dynamical systems on irregular, non-Euclidean domains for nonlinear optimization. SIAM J. Control Optim. **59**(1), 635–668 (2021)
17. Jongen, H.T., Jonker, P., Twilt, F.: Nonlinear Optimization in Finite Dimensions: Morse Theory, Chebyshev Approximation, Transversality, Flows, Parametric Aspects, vol. 47. Springer Science & Business Media (2013)
18. Bhat, S.P., Bernstein, D.S.: Lyapunov analysis of finite-time differential equations. In: Proceedings of 1995 American Control Conference-ACC'95, vol. 3, pp. 1831–1832. IEEE (1995)
19. Haddad, W.M., L'Afflitto, A.: Finite-time stabilization and optimal feedback control. IEEE Trans. Autom. Control **61**(4), 1069–1074 (2015)
20. Prasun, P., Pandey, S., Kamal, S., Ghosh, S., Singh, D.: Discrete-time gradient systems governed by difference equation with minima. In: 2023 31st Mediterranean Conference on Control and Automation (MED), pp. 783–788. IEEE (2023)
21. Kamal, S., Prasun, P.: A Minimum Function-Based Discrete Sliding Mode Control: Theory and Applications. Springer (2026)
22. Polyakov, A.: Nonlinear feedback design for fixed-time stabilization of linear control systems. IEEE Trans. Autom. Control **57**(8), 2106–2110 (2011)
23. Pandey, S., Kamal, S., Osinenko, P., Parsegov, S., Singh, D.: Finite and fixed-time stabilization of discrete-time systems using passivity-based control. Int. J. Robust Nonlinear Control **34**(12), 8481–8494 (2024)
24. Sánchez-Torres, J.D., Gómez-Gutiérrez, D., López, E., Loukianov, A.G.: A class of predefined-time stable dynamical systems. IMA J. Math. Control Inf. **35**(Supplement_1), i1–i29 (2018)
25. Sánchez-Torres, J.D., Muñoz-Vázquez, A.J., Defoort, M., Jiménez-Rodríguez, E., Loukianov, A.G.: A class of predefined-time controllers for uncertain second-order systems. Eur. J. Control **53**, 52–58 (2020)
26. Pal, A.K., Kamal, S., Nagar, S.K., Bandyopadhyay, B., Fridman, L.: Design of controllers with arbitrary convergence time. Automatica **112**, 108710 (2020)
27. Pal, A.K., Kamal, S., Yu, X., Nagar, S.K., Xiong, X.: Free-will arbitrary time consensus for multiagent systems. IEEE Trans. Cybern. **52**, 4636 (2020)
28. Pal, A.K., Kamal, S., Yu, X., Nagar, S.K., Bandyopadhyay, B.: Free-will arbitrary time terminal sliding mode control. IEEE Trans. Circuits Syst. II Express Briefs (2022)

29. Cui, L., Jiang, Z.-P., Sontag, E.D.: Small-disturbance input-to-state stability of perturbed gradient flows: Applications to LQR problem. Syst. Control Lett. **188**, 105804 (2024)
30. Kamal, S., Pal, A.K., Singh, B., Bandyopadhyay, B., Yu, X.: Stabilization of continuous time uncertain systems with predefined upper bound of settling time. In: 2022 16th International Workshop on Variable Structure Systems (VSS), pp. 266–271. IEEE (2022)
31. Cortés, J.: Finite-time convergent gradient flows with applications to network consensus. Automatica **42**(11), 1993–2000 (2006)
32. Prasun, P., Pandey, S., Kamal, S., Ghosh, S., Singh, D., Ghosh, D.: Predefined upper bound of settling time based convergent gradient flow systems. In: IECON 2022–48th Annual Conference of the IEEE Industrial Electronics Society, pp. 1–6. IEEE (2022)
33. Pandey, S., Prasun, P., Kamal, S., Singh, D., Dinh, T.N.: Optimisation by predefined upper bound of settling time convergent projected gradient flows. Int. J. Syst. Sci., 1–14 (2025)
34. Dall'Anese, E., Simonetto, A., Becker, S., Madden, L.: Optimization and learning with information streams: Time-varying algorithms and applications. IEEE Signal Process. Mag. **37**(3), 71–83 (2020)
35. Hazan, E., et al.: Introduction to online convex optimization. Found. Trends® Optim. **2**(3–4), 157–325 (2016)
36. Simonetto, A., Dall'Anese, E., Paternain, S., Leus, G., Giannakis, G.B.: Time-varying convex optimization: Time-structured algorithms and applications. Proc. IEEE **108**(11), 2032–2048 (2020)
37. Besbes, O., Gur, Y., Zeevi, A.: Non-stationary stochastic optimization. Oper. Res. **63**(5), 1227–1244 (2015)
38. Lesage-Landry, A., Shames, I., Taylor, J.A.: Predictive online convex optimization. Automatica **113**, 108771 (2020)
39. Fattahi, S., Josz, C., Mohammadi, R., Lavaei, J., Sojoudi, S.: Absence of spurious local trajectories in time-varying optimization: A control-theoretic perspective. In: 2020 IEEE Conference on Control Technology and Applications (CCTA), pp. 140–147. IEEE (2020)
40. Rahili, S., Ren, W.: Distributed continuous-time convex optimization with time-varying cost functions. IEEE Trans. Autom. Control **62**(4), 1590–1605 (2016)
41. Pandey, S., Prasun, P., Kamal, S., Singh, D., Ghosh, D., Olaru, S.: Optimal trajectory tracking using Newton and Levenberg-Marquardt-like algorithms for constrained time-varying optimization. Optim. Control Appl. Methods **46**(4), 1800–1813 (2025)
42. Hauswirth, A., Bolognani, S., Hug, G., Dörfler, F.: Timescale separation in autonomous optimization. IEEE Trans. Autom. Control **66**(2), 611–624 (2020)
43. Krstić, M., Kanellakopoulos, I., Kokotović, P.: Adaptive nonlinear control without overparametrization. Syst. Control Lett. **19**(3), 177–185 (1992)

44. Ariyur, K.B., Krstic, M.: Real-time Optimization by Extremum-seeking Control. John Wiley & Sons (2003)
45. Diana, B., Pandey, S., Kamal, S., Dinh, T.N.: Finite and fixed-time feedback-based continuous-time optimization. Automatica **183**, 112569 (2026)

Varanasi, India
April, 2026

Shyam Kamal
Sunidhi Pandey

Acknowledgments Although our names appear on the cover, we remain deeply conscious that this work is the culmination of the support, guidance, and contributions of many remarkable individuals. It would not have been possible without their encouragement and generosity of spirit. We extend our sincere gratitude to all who have, in various ways, shaped this journey into a meaningful and enriching experience, one that we will always cherish. We offer our deepest appreciation and respect to the members of the Department of Electrical Engineering at IIT (BHU), Varanasi, whose constant guidance, thoughtful oversight, and unwavering encouragement have been instrumental throughout this endeavor. Their support has played a defining role in bringing this work to completion. In particular, we express our heartfelt gratitude to Prof. Devender Singh for his invaluable guidance, insightful direction, and steadfast support at every stage of this journey.

We would like to express our sincere gratitude to Prof. Sorin Olaru from CentraleSupélec and a member of the CNRS Laboratory of Signals and Systems, Université Paris-Saclay, France, for his valuable insights and constructive ideas in the development of several manuscripts related to this work. We are also deeply thankful to Dr. Debdas Ghosh from the Department of Mathematical Sciences, IIT (BHU) Varanasi, for his insightful mathematical inputs that have significantly strengthened the theoretical aspects of this research. Furthermore, We extend our appreciation to Dr. Thach N. Dinh, Associate Professor at Conservatoire National des Arts et Métiers, France, for his collaborative support, thoughtful discussions, and valuable guidance, which have greatly contributed to enhancing the overall quality of this work.

Finally, and of deepest importance, we express our heartfelt gratitude to our parents and family members for their unwavering support, constant encouragement, and enduring belief in us throughout this journey. Their steadfast confidence in our abilities has been a constant source of strength, especially in times of challenge and uncertainty.

Contents

Acronyms

Abbreviations

GF	Gradient flows
LLS	Linear least squares
PGF	Projected gradient flows
RAP	Resource allocation problem
FTS	Finite-time stable
ISS	Input-to-state stable
FTISS	Finite-time input-to-state stable
UBST	Upper bound of settling time
TVO	Time-varying optimization
KKT	Karush-Kuhn-Tucker
P-L	Polyak- Lojasiewicz
FxTS	Fixed-time stable
SD-ISS	Small-disturbance input to state stability
LEC	Linear equality constraints
min	Minimum
max	Maximum

List of Symbols

$\mathbb{R}$	Real number set
$\mathbb{R}_{\geq 0}$	Set of non-negative real numbers
$\mathbb{R}_{+}$	Set of positive real numbers
$\mathbb{R}^n$	An n-dimensional vector space over the field of real numbers $\mathbb{R}$
$\mathcal{K}$	Class $\mathcal{K}$ function
$\mathcal{GK}$	Generalized-$\mathcal{K}$ function
$\mathcal{L}$	Class $\mathcal{L}$ function
$\mathcal{K}_{\infty}$	Class $\mathcal{K}_{\infty}$ function
$\mathcal{KL}$	Class $\mathcal{KL}$ function
$\mathcal{GKL}$	Generalized-$\mathcal{KL}$ class function

Chapter 1
Introduction

Optimization, in its most fundamental sense, is concerned with identifying the most suitable decision in accordance with a prescribed performance criterion. Classical formulations typically cast such problems in finite-dimensional settings, where the objective is to minimize (or maximize) a function over a set of admissible candidates. In recent years, however, a complementary and increasingly influential viewpoint has emerged—one that interprets optimization through the framework of dynamical systems evolving over continuous time (CT). From this perspective, the process of optimization is not merely a static computation, but a dynamic evolution, where the solution is approached progressively as the system trajectories unfold over time.

A central element in continuous-time optimization (CTO) is the use of a dynamical system to describe how decision variables evolve over time. In this framework, the solution is reached asymptotically or, in some cases, in finite time, through the evolution of the system. Such dynamics are typically expressed in terms of differential equations of the form

$$\dot{x} = \mathcal{G}(x(t))$$

where $x(t) \in \mathbb{R}^n$ represents the decision variable at time t. The vector field $\mathcal{G}$ is designed so that the equilibrium points of the system coincide with the solutions of the underlying optimization problem, thereby linking the notion of optimality with the stability properties of the dynamical system.

Another fundamental aspect of CTO is the objective function, which encodes the performance criterion to be minimized. In contrast to classical static formulations, where the objective serves only as a measure of optimality, here it assumes a more influential role. It not only defines the optimization problem but also actively shapes the dynamics that guide the system toward an optimal solution. In many formulations, the governing vector field $\mathcal{F}$ is constructed directly from the objective function itself—for example, through gradient-based mechanisms or,

S. Kamal, S. Pandey, *A Continuous-Time Perspective on Accelerated Optimization and Control*, Studies in Systems, Decision and Control 693,
https://doi.org/10.1007/978-3-032-31645-5_1

more generally, via operator-theoretic approaches. This close interplay ensures that the evolution of the system is inherently aligned with the structure of the objective, enabling the trajectories to progressively move toward optimality.

From this perspective, a CTO problem may be informally described as the task of designing a dynamical system whose trajectories evolve toward the set of minimizers of a given objective function, either asymptotically or within finite time. In doing so, the notion of optimality is naturally embedded within the long-term behavior of the system. This viewpoint fundamentally reshapes the study of optimization, recasting it as a problem of stability and convergence analysis. As a result, the tools of differential equations, dynamical systems, and control theory become indispensable, providing a rigorous framework for understanding how and why such systems converge to optimal solutions.

It is important to recognize that optimality is not merely an abstract mathematical construct, but a fundamental principle that pervades both natural phenomena and engineered systems. Many physical processes inherently evolve toward configurations that minimize energy, while engineered systems are deliberately designed to achieve efficiency, robustness, and performance. In both cases, optimization principles often manifest implicitly through the underlying dynamics. CTO provides a systematic and rigorous framework to formalize, analyze, and exploit this intrinsic connection between dynamical evolution and optimal behavior.

In this monograph, our emphasis is placed on the mathematical foundations and system-theoretic perspective of CTO. Rather than focusing on specific applications, the objective is to develop a unified and coherent framework that elucidates how optimization problems can be interpreted, analyzed, and solved using tools from dynamical systems, stability theory, and control.

To build intuition and rigor, we will first revisit the basics of finite-dimensional optimization. This will allow us to clearly distinguish between static optimality conditions and their dynamic realizations. Once these foundations are established, we will return to CT formulations and explore how optimization algorithms can be interpreted, analyzed, and designed as dynamical systems.

The CT formulation of optimization problems enables simple and elegant convergence proofs to equilibrium points, leveraging control theory tools such as Lyapunov stability theory. First-order iterative algorithms are widely used in machine learning to handle large-scale datasets and leverage parallel processing architectures. Gradient-based methods constitute a class of optimization algorithms recognized for their robustness to measurement noise, computational simplicity, and modest memory requirements [1]. A classic example is the gradient flow (GF), defined by

$$\dot{x} = -\nabla g(x),$$

where $g : \mathbb{R}^n \to \mathbb{R}$ is a continuously differentiable, unconstrained objective function that we aim to minimize. It is well established that strict local minimizers of a locally convex function $g : \mathbb{R}^n \to \mathbb{R}$ correspond to stable equilibria of the gradient flow (GF) dynamics. Moreover, if the sublevel sets of g are compact, then all trajectories converge asymptotically to the set of critical points of g.

Classical techniques include dual methods, Lagrangian multiplier techniques, penalty (barrier) function approaches, and projected gradient techniques to solve constrained optimization problems. The optimization problems, including inequality constraints, are handled by extending first-order conditions to Karush-Kuhn-Tucker (KKT) conditions.

The approaches discussed above for solving optimization problems generally assume a static setting, where neither the objective function nor the constraints change over time. However, in many real-time applications, optimization problems are dynamic—meaning the objective function or constraints evolve based on certain parameters. In time-varying optimization (TVO) problems, either the objective or the constraints are explicitly time-dependent and change continuously over time, leading to an optimizer trajectory composed of the optimal solutions at each moment.

A traditional approach to solving TVO problems involves sampling the objective and constraint functions at specific time instants and solving the resulting optimization problem at each of these points. This method demands significant computational effort and ensures optimality provided that each sampled problem converges within a time interval shorter than the sampling period. An alternate useful approach is based on time-structured algorithms. Time-structured algorithms exploit the intrinsic properties of the optimization problem, such as differentiability and Lipschitz continuity. A widely used time-structured approach is the prediction-correction method, which consists of a "prediction" step to estimate the optimizer trajectory at the next time instant, followed by a "correction" step to adjust the prediction toward the actual optimizer path. However, this algorithm typically relies on the strong convexity of the objective function to effectively handle both constrained and unconstrained time-varying optimization problems.

The perspective of CTO not only provides valuable insights and concepts for analyzing and developing algorithms but also offers intuitive methods for designing control strategies for physical systems. In contrast to traditional approaches that compute optimal solutions offline and implement them as set-points, the autonomous optimization methodology embeds optimization dynamics directly within the system evolution, thereby enabling real-time, closed-loop optimization. Such an approach leverages the interpretation of optimization algorithms as dynamical systems and as feedback controllers interacting with physical processes, allowing the system to autonomously converge to optimal operating points using real-time information [2]. In particular, gradient-based feedback laws can be systematically extended to non-Lipschitz and dual-exponent structures to achieve accelerated convergence guarantees, including finite-time and fixed-time stability (FxTS). The resulting framework not only enhances convergence speed but also improves robustness to initial conditions and disturbances, thereby providing a principled foundation for fast and reliable optimization-driven control of nonlinear dynamical systems.

1.1 Accelerated Continuous-Time Optimization: A Paradigm Shift

The evolution of optimization theory has been driven by the persistent need for faster convergence, improved robustness, and deeper structural understanding of complex problems. Classical gradient-based methods provided the first systematic framework for iterative minimization. However, these methods often suffer from slow convergence, particularly in ill-conditioned settings. A major breakthrough came with the introduction of acceleration techniques by Yurii Nesterov [3], which significantly improved convergence rates and reshaped modern optimization theory.

A fundamental shift occurred when accelerated methods were interpreted through the lens of CT dynamical systems. In this framework, optimization algorithms are viewed as discretizations of differential equations. This perspective was formalized in the seminal work [4], where Nesterov's accelerated gradient method was shown to correspond to a second-order ordinary differential equation with time-varying damping.

In classical CTO, the GF is described by the first-order system

$$\dot{x}(t) = -\nabla f(x(t)), \tag{1.1}$$

which guarantees monotonic decrease of the objective function but exhibits relatively slow convergence. In contrast, accelerated dynamics introduce an inertial term, leading to a second-order system of the form

$$\ddot{x}(t) + \frac{\alpha}{t}\dot{x}(t) + \nabla f(x(t)) = 0, \tag{1.2}$$

where the time-dependent damping term plays a crucial role in achieving faster convergence. This formulation reveals that acceleration is not merely an algorithmic artifact but an intrinsic property of certain dynamical systems.

This continuous-time viewpoint enables a paradigm shift in algorithm design. Instead of directly constructing discrete-time iterations, one can first design a dynamical system with desired convergence properties—such as exponential, finite-time, or fixed-time convergence—and subsequently discretize it to obtain implementable algorithms. This approach naturally aligns with tools from control theory, including Lyapunov stability, and dissipativity.

An important insight emerging from this perspective is the energy-based interpretation of accelerated methods. The dynamics can often be associated with a Lyapunov function resembling the total energy of a mechanical system, combining potential energy (objective function) and kinetic energy (velocity term). The systematic decay of this energy along trajectories provides a rigorous certificate of convergence. Furthermore, recent advances have extended this viewpoint to contraction analysis and Riemannian metrics, offering a geometric interpretation of optimization dynamics.

The paradigm also facilitates the unification of several optimization methods. Polyak's heavy-ball method [5], Nesterov's acceleration [3], and variational formulations [6] can all be interpreted as specific instances of second-order or higher-order dynamical systems. This unified framework not only clarifies the relationships among these methods but also enables systematic development of new algorithms with improved performance guarantees.

In summary, accelerated continuous-time optimization represents a fundamental shift in perspective—from viewing algorithms as discrete procedures to understanding them as dynamical systems. This paradigm not only deepens theoretical insights but also offers a principled framework for designing faster, more robust, and more scalable optimization methods, thereby bridging the gap between optimization and control theory.

1.2 Motivation

Gradient flow is one of the time-honored and famous results of continuous-time optimization. The standard gradient flow system is asymptotically or exponentially convergent, meaning that a strict optimal solution is obtained when the time approaches infinity. Asymptotic/exponential convergence of the solution to the optimal point is not desirable for time-critical tasks or multi-cycle optimization.

It is well known that since the introduction of finite-time stability (FTS), the system and control community has put a great deal of effort into obtaining less conservative convergence, resulting in many new concepts and results such as FTS/FxTS. Then, a problem naturally arises: is it possible to design more efficient gradient flow systems with the help of more advanced stability theory? Thus, the gradient dynamical systems ensuring finite-time or fixed-time convergence (FxTC) have received a lot of attention. Normalized and signed gradient flows guarantee finite-time convergence (FTC), where the upper bound of settling time (UBST) is related to the initial gradient value. Fixed-time gradient flows guarantee fixed-time convergence, where the upper bound of settling time is uniform irrespective of the initial conditions. The upper bound of the settling time is often difficult to specify, either depending on system initialization or involving complex parameter calculation. One of the motivations of this monograph is to provide an answer to the following question: how could one modify the GF dynamics so that the trajectories converge to the critical points of the function in a prior chosen time?

The main motivation of this monograph is to explore the interplay between optimization and control theory. Control theory provides a rich and well-developed framework, offering a wide range of mathematical tools. In this context, it can be employed to ensure algorithmic convergence to accurate solutions and to systematically analyze the effects of numerical errors and computational delays. Lyapunov stability theory provides elegant and simple proofs for the convergence of solutions to optimal points. Control-theoretic tools have been widely applied in the context of time-invariant optimization problems. This monograph uses the tools of control theory to develop algorithms to solve TVO problems with equality and inequality constraints. Further, advanced studies related to Lyapunov stability theory are used in convergence analysis. This monograph also investigates the applicability of the developed novel approaches to solve TVO problems in a real-world example like the navigation of a robot in an environment with circular obstacles.

Motivated by these observations, it is natural to move beyond purely algorithmic gradient flows and investigate optimization dynamics in a feedback setting, where the evolution of the decision variables is inherently coupled with the dynamics of physical systems. In many practical scenarios, the optimization variable is not an abstract quantity but is directly linked to system inputs, and the objective function depends on the system state. This raises a fundamental question: can one design optimization dynamics that operate in closed loop with the plant while still preserving strong convergence guarantees such as finite-time or fixed-time stability?

In this direction, embedding optimization within feedback loops provides a powerful mechanism to achieve real-time optimal behavior. Unlike conventional approaches that compute optimal solutions offline and subsequently apply them to the system, feedback-based optimization allows the system to continuously adjust its inputs based on instantaneous measurements. However, the introduction of system dynamics also complicates the analysis, as the convergence of the optimization variables must now be studied jointly with the stability of the plant.

To address these challenges, one can exploit advanced Lyapunov-based techniques to construct composite energy-like functions that capture both the optimization error and the system dynamics. This naturally leads to the design of modified gradient flows, where non-Lipschitz and multi-term structures are introduced to accelerate convergence. In particular, finite-time feedback laws enable faster convergence near the optimal point, while fixed-time designs ensure that convergence occurs within a prescribed bound independent of initial conditions.

These developments highlight an important shift in perspective: optimization is no longer treated as a standalone computational task, but as an integral part of the system dynamics. This viewpoint not only provides deeper insight into the behavior of optimization algorithms but also enables the systematic design of fast and robust control strategies for nonlinear systems.

1.3 Related Work

CTO is currently a vibrant area of research within optimization theory, with prior studies offering valuable insights and elegant techniques for establishing stability and convergence properties. This literature review provides a comprehensive overview of existing research in the field of continuous-time optimization, encompassing various gradient flows, unconstrained and constrained time-varying optimization problem-solving techniques, analysis of system properties using gradient flows, and related areas. It seeks to highlight the key contributions and emerging trends within this evolving field.

1.3.1 Gradient Flows

Guidelines for designing dynamical systems for optimization, with particular emphasis on gradient-based dynamics, are presented in [7]. Moreover, a substantial body of literature exists on the asymptotic convergence analysis of gradient flows [4, 6]. In [4], the focus is on the exponential stability of the GF-based methods. The strong or strict convexity of the objective function is a standard assumption. In [8], it is shown that the condition can be relaxed by assuming that the objective function satisfies the Polyak–Lojasiewicz (PL) inequality, i.e., the objective function is gradient dominated. The following Lemma summarizes the classical results related to gradient flow present in the literature.

Lemma 1.1 ([9–11]) *Let $f : \mathbb{R}^n \to \mathbb{R}$, be continuously differentiable whose gradient $\nabla f : \mathbb{R}^n \to \mathbb{R}^n$ is locally Lipschitz. Assume that for some $m \in \mathbb{R}$, the sublevel set*

$$f^{-1}(m) = \{\zeta \in \mathbb{R}^n | f(\zeta) \leq m\}$$

is compact. Then, for the GF

$$\dot{\zeta} = -\nabla f(\zeta)$$

following statements hold:

1. *Any trajectory with initial condition in $f^{-1}(m)$ converges to the set of critical points, i.e., points ζ^* satisfying $\nabla f(\zeta^*) = 0$.*
2. *If f is analytic or convex, then every such trajectory starting in $f^{-1}(m)$ converges to a single point.*
3. *Every asymptotically stable equilibrium is a strict and isolated minimizer, and every local minimizer is stable. If f is analytic or convex, then the set of all (strict) minimizers is equivalent to the set of (asymptotically) stable equilibria.*

1.3.1.1 Unconstrained Optimization: Finite-Time/Fixed-Time Convergent Gradient Flows

Consider the following unconstrained minimization problem

$$\min_{\zeta \in \mathbb{R}^n} g(\zeta) \tag{1.3}$$

where $g : \mathbb{R}^n \to \mathbb{R}$. The following assumption is necessary for the convergence of gradient-based methods to an optimal solution.

Assumption 1 *There exists $\zeta^* \in \mathbb{R}^n$ s.t. $-\infty < g^* = g(\zeta^*)$.*

Lemma 1.2 ([12]) *If g is convex and continuously differentiable, then a point ζ^* is a global minimizer of g if and only if $\nabla g(\zeta^*) = 0$. Moreover, if g is strictly convex, then this minimizer ζ^* is unique.*

In [13], the authors introduce the normalized and signed gradient flows associated with a differentiable function that have the property of convergence to critical points of a function in finite time. In [6], the authors discuss a rescaled accelerated gradient flow dynamics having the following dynamics:

$$\dot{\zeta} = -\frac{\nabla g(\zeta)}{\|g(\zeta)\|^{\frac{q-2}{q-1}}}, \tag{1.4}$$

where, $q > 2$.

Lemma 1.3 ([6]) *If $g : \mathbb{R}^n \to \mathbb{R}$ is continuously differentiable and strongly convex, then the trajectories of* (1.4) *converge to the optimal point ζ^* in finite time, for $q > 2$.*

Inspired from (1.4), authors in [14] modified GF dynamics, so that the optimal point of (1.3) can be obtained within a fixed time for any $\zeta(0) \in \mathbb{R}^n$. In [14], the authors propose the following fixed-time convergent GF approach:

$$\dot{\zeta} = -c_1 \frac{\nabla g(\zeta)}{\|\nabla g(\zeta)\|^{\frac{q_1-2}{q_1-1}}} - c_2 \frac{\nabla g(\zeta)}{\|\nabla g(\zeta)\|^{\frac{q_2-2}{q_2-1}}} \tag{1.5}$$

where, $q_1 > 2$, $1 < q_2 < 2$, and $c_1, c_2 > 0$.

Lemma 1.4 ([14]) *Under Assumption 1 on the objective function g, the trajectories of* (1.5) *reach the optimal point ζ^* within a fixed time, independent of the initial condition $\zeta(0)$.*

1.3.1.2 Convex Optimization With Linear Equality Constraints: Finite-Time/Fixed-Time Convergent Gradient Flows

Consider the following constrained optimization problem

$$\min_{\zeta \in \mathbb{R}^n} g(\zeta)$$

$$\text{subject to } C\zeta = d \tag{1.6}$$

where, $\zeta \in \mathbb{R}^n$, $g : \mathbb{R}^n \to \mathbb{R}$ is convex and continuously differentiable, $C \in \mathbb{R}^{m \times n}$, and $d \in \mathbb{R}^m$. In [1], the authors propose the following signed projected gradient method to solve the optimization problem (1.6):

$$\dot{\zeta} = -\beta P^\top \text{sign}(P \nabla g(\zeta)), \tag{1.7}$$

where, $P \in \mathbb{R}^{n \times n}$ is called projection matrix, sign$(\cdot)$ is the signum function defined componentwise, and $\beta > 0$. Following the assumption that columns of matrix C are linearly independent, authors in [1] discuss the FTC analysis of the sign projected gradient system (1.7).

1.3.2 Time-Varying Optimization

Most optimization problems encountered in real-time applications—such as guidance and navigation of aerial and autonomous vehicles, machine learning, and economics—are dynamic in nature [15–17]. A TVO problem involves an objective function and/or constraints that depend on time. Gradient flows are modified to solve TVO problems. To address this problem in a practical setting, an algorithm should have the following two properties: convergence to the time-varying optimum and the ability to track it. In [18], the author shows that the gradient method is able to converge to an arbitrarily small neighborhood of the optimum, assuming strong convexity and twice differentiability of the objective function. In [19], a dynamics is proposed that is able to converge, in the sense that $\lim_{\zeta \to \infty} \frac{\partial g(t,\zeta)}{\partial t} = 0$, assuming objective function $g(t, \zeta)$ is twice differentiable with respect to ζ for all t. In the literature, commonly used techniques to solve TVO problems are online optimization [20, 21], batch algorithms [22], and prediction—correction methodology. Online optimization is an unstructured algorithm [22]. It computes the optimal points at each time instant based solely on past information, without utilizing the inherent properties of the underlying optimization problem. In Batch algorithms, objective functions and constraints are sampled, and at each instant corresponding sequence of time-invariant constrained optimization problems are solved. This approach ignores the dynamic aspect of the problem, since each iteration tends to converge toward the optimal point of the sampled time-invariant problem, while the solution of the TVO problem is drifting away over time. Whereas, the prediction-correction methodology is a time-structured approach.

Prediction—correction algorithms are commonly employed to track the optimizer trajectories of time-varying convex optimization problems. The prediction correction algorithm assumes strong convexity of the objective function to solve unconstrained [23] and constrained [24] TVO problems. These methods consist of a "prediction" step to estimate the optimizer trajectory at the next time instant, followed by a "correction" step to adjust the predicted value toward the true optimizer path. A prediction—correction-based CT dynamical system approach to solve unconstrained TVO is proposed in [25, 26], and guarantees the asymptotic convergence of the trajectories of the proposed system to the optimizer trajectory. More precisely, they propose a CT dynamical system

$$\dot{\zeta} = f(t, \zeta(t))$$

whose solution $\zeta(t)$ satisfies $\|\zeta(t)-\zeta^*(t)\| \to 0$ as $t \to \infty$ for all initial conditions. In [26], a projected dynamical system framework is proposed to track the optimizer trajectory of a time-varying nonconvex optimization problem. Within this approach, inequality constraints are addressed by projecting the iterates onto the feasible set at each time instant. Additionally, inequality-constrained TVO problems are treated using logarithmic barrier functions with suitably selected slack variables and barrier parameters.

In [27], the authors develop a projected dynamical system approach for tracking the optimizer trajectory of a convex optimization problem with time-varying objective functions and inequality constraints. This method is constructed based on the Karush–Kuhn–Tucker (KKT) optimality conditions, along with an explicit characterization of the optimizer trajectory derived from these conditions. The asymptotic as well as FxTC of the proposed dynamics to the optimizer trajectory is established via Lyapunov-based analysis.

Furthermore, in [28], a projected dynamical system framework is introduced for tracking the optimizer trajectory of a time-varying nonconvex optimization problem, wherein inequality constraints are handled through projection onto the feasible region at each time instant.

1.3.3 Feedback-Based Optimization

Optimization has traditionally been approached as a numerical and algorithmic discipline, where problems are solved offline, and the resulting solutions are implemented as set-points for physical systems [29, 30]. However, such a separation between optimization and system dynamics often leads to suboptimal performance in the presence of modeling uncertainties, disturbances, and time-varying environments [2]. The paradigm of feedback-based optimization (also referred to as autonomous optimization) has emerged as a powerful framework [2, 31]. In contrast to offline optimization, this approach embeds optimization dynamics directly within feedback loops of physical systems, enabling real-time adaptation using measurements. Such methods can be interpreted either as optimization algorithms interacting with system dynamics or as feedback controllers that mimic optimization processes [2]. Early developments in this direction can be traced back to applications in communication networks [32] and power systems [33], where control mechanisms were shown to inherently solve optimization problems.

A central challenge in FBO is the stability of the closed-loop interconnection between the plant and the optimization dynamics. Unlike purely algorithmic settings, the coupling between system dynamics and optimization updates can lead to instability or degraded convergence. To address this issue, the concept of timescale separation has been extensively studied [31]. In particular, it has been shown that stability can be ensured when the plant dynamics evolve on a faster timescale than the optimization dynamics, allowing the system to be approximated by its steady-state behavior. Singular perturbation theory provides a rigorous framework for analyzing such systems and deriving explicit stability conditions [34].

Related research directions include extremum seeking control [35], and real-time optimization. While these approaches share similarities with feedback-based optimization, they differ in their reliance on probing signals, iterative numerical solvers, or model-based predictions. In contrast, feedback-based optimization directly exploits system dynamics to achieve optimal steady-state behavior without explicitly solving the optimization problem at each step [2].

Despite significant progress, most existing works focus on asymptotic or exponential convergence. Such convergence rates may be inadequate for time-critical applications, motivating the development of finite-time and fixed-time optimization methods. These approaches guarantee convergence within finite or uniformly bounded time, offering improved performance and robustness. However, their integration within feedback-based optimization frameworks remains relatively unexplored and constitutes an important direction for further research.

1.4 Summary of Contributions

This monograph constitutes a significant advancement in the field of CTO, introducing a novel framework based on dynamical systems with predefined-time convergent GF. The impact of this work is multifaceted, with its principal contributions reflected in the following aspects:

- At first, our investigation focused on continuous-time unconstrained optimization problems, leading to the development of a novel gradient flow with predefined-time convergence. This flow is rigorously designed for strongly convex objective functions and further extended to accommodate functions satisfying the more relaxed Polyak-Lojasiewicz (P-L) condition. Unlike existing methods, our scheme allows the selection of the upper bound on convergence time a priori—independent of the initial conditions or intricate parameter tuning. This significant simplification facilitates practical implementation while ensuring theoretical guarantees of convergence via Lyapunov-based analysis. Building upon this foundation, we integrated the concept of predefined-time stable gradient flows with projected gradient techniques to address convex optimization problems subject to linear equality constraints. As a result, we proposed a projected gradient flow algorithm with predefined-time convergence, offering the same initialization-free time-setting advantage. Furthermore, to address system perturbations, a perturbed version of the projected flow was analyzed within the input-to-state stability framework, enhancing robustness. The efficacy and practical relevance of the proposed algorithms were validated through numerical studies on linear least squares and resource allocation problems, with simulation results confirming their rapid and reliable convergence performance.

- Next, we extended the discussion on predefined-time gradient flows by involving the partial derivative of the gradient with respect to time, to solve TVO problems. We introduce a novel continuous-time dynamical system capable of tracking optimal trajectories for TVO problems subject to time-varying constraints, achieving convergence within a user-defined, predefined time frame. To address practical limitations, we develop a robust Newton-like algorithm that handles scenarios where the exact rate of change of the objective function's gradient is unavailable. Furthermore, for cases involving singular or near-singular Hessian matrices, we propose a Levenberg-Marquardt-like algorithm, ensuring reliable performance. Rigorous Lyapunov-based convergence analysis underpins the theoretical guarantees for both proposed algorithms. Comprehensive simulation studies demonstrate the efficacy and robustness of this predefined-time approach across diverse time-varying optimization problems.
- Further, this monograph delved into the demonstration of the proposed dynamics in real-world problems like collision-free robot navigation in complex sphere world environments. We develop a vector field-based navigation strategy defined over a compact, convex workspace punctured by disk-shaped obstacles. This strategy guarantees predefined-time convergence to a moving target position for a disk-shaped robot, starting from almost all initial conditions (excluding a set of measure zero) while ensuring strict collision avoidance throughout the trajectory. Crucially, we demonstrate that this navigation problem can be effectively formulated and solved as a TVO problem with inequality constraints. The efficacy and robustness of our proposed predefined-time convergent dynamics, along with their application to this challenging navigation task, are rigorously validated through comprehensive simulation results.
- Finally, our focus shifted to the development of a feedback-based CTO framework with guaranteed FTC and FxTC. In contrast to classical GF that ensures only asymptotic convergence, the proposed approach systematically incorporates non-Lipschitz and dual-exponent feedback structures to achieve accelerated convergence with explicit bounds on settling time. A unified Lyapunov-based analysis is presented through the construction of composite Lyapunov functions, capturing the interaction between plant dynamics and optimization dynamics in a closed-loop setting. The proposed methodology enables real-time optimization by embedding the optimization process within system feedback, thereby eliminating the need for offline computation. Furthermore, the effectiveness of the developed framework is demonstrated through a nonlinear coupled-tank system, highlighting improved convergence speed and robustness. Overall, we provide a systematic control-theoretic approach for designing fast and reliable optimization dynamics for nonlinear systems.

1.5 Structure of the Monograph

This monograph is organized into eight chapters.

The opening chapter, Chap. 1, provides a comprehensive introduction to the work presented. It encompasses a detailed discussion of the underlying motivation, an extensive review of the existing literature to identify prevailing research gaps, and a clear articulation of the principal contributions of the monograph. This chapter lays the groundwork for the subsequent developments, while systematically defining the research objectives. In addition, it highlights the significance and broader relevance of the proposed work within the contemporary academic and research landscape.

Chapter 2 introduces the fundamentals of finite-dimensional optimization, providing the necessary analytical and geometric foundation for the developments in subsequent chapters. It begins with formal definitions of optimality, including local, strict, and global minima, and distinguishes between constrained and unconstrained optimization problems. Classical first- and second-order optimality conditions are derived for unconstrained settings and then extended to constrained problems through the characterization of feasible sets and tangent spaces, leading to Lagrange multiplier-based conditions. The chapter further explores convex sets and convex functions, emphasizing their role in guaranteeing global optimality, and introduces key notions such as strict and strong convexity. Additionally, the PL inequality is discussed as a useful condition for establishing convergence properties beyond convex settings. Overall, this chapter establishes the core tools required for analyzing optimization problems and lays the groundwork for their continuous-time dynamical interpretations in later chapters.

Chapter 3 provides a comprehensive overview of the mathematical preliminaries required for the developments presented in this thesis. It introduces the key notions, definitions, and concepts that form the foundation for understanding the subsequent chapters. In particular, this chapter covers essential topics such as comparison functions and stability concepts within the CT systems framework. By establishing these fundamental elements, it lays the necessary groundwork for the advanced methodologies and theoretical results developed in the remainder of the monograph.

Chapter 4 introduces dynamical systems designed to solve unconstrained and equality-constrained optimization problems, ensuring convergence within a predefined time under the assumption that the objective functions satisfy the gradient dominance condition. The convergence analysis is carried out using Lyapunov-based methods. This chapter also examines the impact of perturbations on the proposed gradient flows through the lens of ISS theory. Furthermore, simulation results on least squares and resource allocation problems demonstrate the effectiveness of the proposed approach.

Chapter 5 presents a class of second-order dynamical systems designed to solve TVO problems with the capability to track the optimal solution trajectory within a predefined time horizon. Initially, a Newton-like dynamic is formulated for optimal trajectory tracking in constrained convex time-varying optimization problems. To address scenarios where the time derivative of the gradient is not explicitly available, a robust Newton-inspired dynamic with predefined-time stability is proposed. Furthermore, to handle situations involving non-invertible or singular Hessians, a Levenberg—Marquardt-type predefined-time stable dynamic is introduced.

Chapter 6 explores the real-world applications of the proposed dynamics in Chap. 5. The proposed approach is applied to robot navigation with collision avoidance, where the robot is required to track a moving target within a predefined time. By formulating the task as a TVO problem with inequality constraints, convergence to the moving target is ensured from almost all initial conditions. Simulation results confirm the robustness and effectiveness of the proposed dynamics in complex, obstacle-rich environments.

Chapter 7 focuses on feedback-based continuous-time optimization with an emphasis on finite-time and fixed-time convergence. It formulates optimization problems within a dynamical systems framework, where the decision variables evolve in a closed loop with the underlying plant dynamics. A unified Lyapunov-based analysis is developed using composite Lyapunov functions to capture the interaction between the accelerated gradient flow dynamics and system dynamics. The theoretical results are further validated through a nonlinear coupled-tank system, demonstrating improved convergence speed and robustness compared to classical approaches.

The final chapter summarizes the principal findings and contributions of the monograph, providing a comprehensive synthesis of the work. It discusses the broader implications of the results, acknowledges the inherent limitations, and outlines promising directions for future research as well as potential practical applications. This chapter serves as the culmination of the study, offering a coherent conclusion to the overall investigation.

References

1. Chen, F., Ren, W.: Sign projected gradient flow: A continuous-time approach to convex optimization with linear equality constraints. Automatica **120**, 109156 (2020)
2. Hauswirth, A., He, Z., Bolognani, S., Hug, G., Dörfler, F.: Optimization algorithms as robust feedback controllers. Annu. Rev. Control **57**, 100941 (2024)
3. Nesterov, Y.: Introductory Lectures on Convex Optimization: A Basic Course, vol. 87. Springer Science & Business Media (2013)
4. Su, W., Boyd, S., Candes, E.J.: A differential equation for modeling nesterov's accelerated gradient method: Theory and insights. J. Mach. Learn. Res. **17**(153), 1–43 (2016)
5. Polyak, B.T.: Some methods of speeding up the convergence of iteration methods. USSR Comput. Math. Math. Phys. **4**(5), 1–17 (1964)
6. Wibisono, A., Wilson, A.C., Jordan, M.I.: A variational perspective on accelerated methods in optimization. Proc. Natl. Acad. Sci. **113**(47), E7351–E7358 (2016)
7. Helmke, U., Moore, J.B.: Optimization and Dynamical Systems. Springer Science & Business Media (2012)
8. Karimi, H., Nutini, J., Schmidt, M.: Linear convergence of gradient and proximal-gradient methods under the Polyak-Łojasiewicz condition. In: Joint European Conference on Machine Learning and Knowledge Discovery in Databases, pp. 795–811. Springer (2016)
9. Absil, P.-A., Kurdyka, K.: On the stable equilibrium points of gradient systems. Syst. Control Lett. **55**(7), 573–577 (2006)
10. Brown, A.A., Bartholomew-Biggs, M.C.: Some effective methods for unconstrained optimization based on the solution of systems of ordinary differential equations. J. Optim. Theory Appl. **62**, 211–224 (1989)

11. Hirsch, M.W., Devaney, R.L., Smale, S.: Differential Equations, Dynamical Systems, and Linear Algebra, vol. 60. Academic Press (1974)
12. Boyd, S.P., Vandenberghe, L.: Convex Optimization. Cambridge University Press (2004)
13. Cortés, J.: Finite-time convergent gradient flows with applications to network consensus. Automatica **42**(11), 1993–2000 (2006)
14. Garg, K., Panagou, D.: Fixed-time stable gradient flows: Applications to continuous-time optimization. IEEE Trans. Autom. Control **66**(5), 2002–2015 (2020)
15. Jin, J., Chen, W., Chen, C., Chen, L., Tang, Z., Chen, L., Wu, L., Zhu, C.: A predefined fixed-time convergence ZNN and its applications to time-varying quadratic programming solving and dual-arm manipulator cooperative trajectory tracking. IEEE Trans. Ind. Inf. **19**(8), 8691–8702 (2022)
16. Dall'Anese, E., Simonetto, A.: Optimal power flow pursuit. IEEE Trans. Smart Grid **9**(2), 942–952 (2016)
17. Dall'Anese, E., Simonetto, A., Becker, S., Madden, L.: Optimization and learning with information streams: Time-varying algorithms and applications. IEEE Signal Process. Mag. **37**(3), 71–83 (2020)
18. Popkov, A.Y.: Gradient methods for nonstationary unconstrained optimization problems. Autom. Remote Control **66**, 883–891 (2005)
19. Zhao, Y., Lu, W.: Training neural networks with time-varying optimization. In: Proceedings of 1993 International Conference on Neural Networks (IJCNN-93-Nagoya, Japan), vol. 2, pp. 1693–1696. IEEE (1993)
20. Hazan, E., et al.: Introduction to online convex optimization. Found. Trends® Optim. **2**(3–4), 157–325 (2016)
21. Colombino, M., Dall'Anese, E., Bernstein, A.: Online optimization as a feedback controller: Stability and tracking. IEEE Trans. Control Netw. Syst. **7**(1), 422–432 (2019)
22. Simonetto, A., Dall'Anese, E., Paternain, S., Leus, G., Giannakis, G.B.: Time-varying convex optimization: Time-structured algorithms and applications. Proc. IEEE **108**(11), 2032–2048 (2020)
23. Simonetto, A., Mokhtari, A., Koppel, A., Leus, G., Ribeiro, A.: A class of prediction-correction methods for time-varying convex optimization. IEEE Trans. Signal Process. **64**(17), 4576–4591 (2016)
24. Simonetto, A.: Dual prediction–correction methods for linearly constrained time-varying convex programs. IEEE Trans. Autom. Control **64**(8), 3355–3361 (2018)
25. Zhao, Y., Swamy, M.: A novel technique for tracking time-varying minimum and its applications. In: Conference Proceedings. IEEE Canadian Conference on Electrical and Computer Engineering (Cat. No. 98TH8341), vol. 2, pp. 910–913. IEEE (1998)
26. Fazlyab, M., Paternain, S., Preciado, V.M., Ribeiro, A.: Prediction-correction interior-point method for time-varying convex optimization. IEEE Trans. Autom. Control **63**(7), 1973–1986 (2017)
27. Raveendran, R., Mahindrakar, A.D., Vaidya, U.: Dynamical system approach for time-varying constrained convex optimization problems. IEEE Trans. Autom. Control **69**(6), 3822–3834 (2023)
28. Hauswirth, A., Subotić, I., Bolognani, S., Hug, G., Dörfler, F.: Time-varying projected dynamical systems with applications to feedback optimization of power systems. In: 2018 IEEE Conference on Decision and Control (CDC), pp. 3258–3263 (2018)
29. Bertsimas, D., Tsitsiklis, J.N.: Introduction to Linear Optimization, vol. 6. Athena Scientific, Belmont (1997)
30. Bertsekas, D.P., Nonlinear programming. J. Oper. Res. Soc. **48**(3), 334–334 (1997)
31. Hauswirth, A., Bolognani, S., Hug, G., Dörfler, F.: Timescale separation in autonomous optimization. IEEE Trans. Autom. Control **66**(2), 611–624 (2020)
32. Low, S.H., Lapsley, D.E.: Optimization flow control. I. Basic algorithm and convergence. IEEE/ACM Trans. Netw. **7**(6), 861–874 (1999)

33. Dörfler, F., Bolognani, S., Simpson-Porco, J.W., Grammatico, S.: Distributed control and optimization for autonomous power grids. In: 2019 18th European Control Conference (ECC), pp. 2436–2453. IEEE (2019)
34. Kokotović, P., Khalil, H.K., O'reilly, J.: Singular Perturbation Methods in Control: Analysis and Design. SIAM, Philadelphia (1999)
35. Ariyur, K.B., Krstic, M.: Real-time Optimization by Extremum-seeking Control. John Wiley & Sons (2003)

Chapter 2
Basics on Finite-Dimensional Optimization

Abstract This chapter provides a comprehensive introduction to finite-dimensional optimization, forming the foundational basis for the continuous-time optimization frameworks developed in subsequent chapters. We begin with the formal definition of optimality, including local, strict, and global minima, and discuss the distinction between constrained and unconstrained optimization problems.For unconstrained settings, first- and second-order necessary and sufficient conditions for optimality are derived using classical tools from multivariable calculus. These results are then extended to constrained optimization problems, where the geometry of feasible sets is characterized through tangent spaces, leading to the formulation of first-order optimality conditions via Lagrange multipliers. The chapter further develops the theory of convex sets and convex functions, highlighting their fundamental role in guaranteeing global optimality. Key concepts such as strict convexity and strong convexity are introduced, along with their equivalent first- and second-order characterizations. In addition, the Polyak–Lojasiewicz (PL) inequality is presented as a weaker condition that still ensures favorable convergence properties, even in certain nonconvex settings. Overall, this chapter establishes the essential analytical and geometric tools required for understanding optimization problems in finite-dimensional spaces, and sets the stage for their dynamic interpretation in continuous-time optimization.

2.1 An Overview of Finite-Dimensional Optimization

Let $f : \mathbb{R}^n \to \mathbb{R}$ be a real-valued function defined on a set $\mathcal{D} \subseteq \mathbb{R}^n$. A point $x^* \in \mathcal{D}$ is said to be a local minimizer of f over $\mathcal{D}$ if there exists a scalar $\epsilon > 0$ such that

$$f(x^*) \leq f(x), \quad \forall x \in \mathcal{D} \cap \mathbb{B}_\epsilon(x^*)$$

where $\mathbb{B}_\epsilon(x^*) := \{x \in \mathbb{R}^n \,|\, \|x - x^*\| < \epsilon\}$ denotes the open ball of radius ϵ centered at x^*. It is important to emphasize that this definition pertains exclusively to points within the set $\mathcal{D}$.

S. Kamal, S. Pandey, *A Continuous-Time Perspective on Accelerated Optimization and Control*, Studies in Systems, Decision and Control 693,
https://doi.org/10.1007/978-3-032-31645-5_2

A point $x^* \in \mathcal{D}$ is termed a strict local minimizer if

$$f(x^*) < f(x), \quad \forall x \in \mathcal{D}, \quad x \neq x^* \text{ with } \|x - x^*\| < \epsilon,$$

and if $f(x^*) \leq f(x), \quad \forall x \in \mathcal{D}$, then x^* is a global minimizer over $\mathcal{D}$.

A minimizer is not necessarily unique; uniqueness is ensured only when it is both strict and global. The notions of maxima are defined analogously, and any point that is either a maximum or a minimum is called an extremum. Since maxima of f are equivalent to minima of $-f$, it is sufficient to focus exclusively on minimization problems.

2.1.1 *Unconstrained Optimization*

The term unconstrained optimization typically refers to settings in which, for a given point $x^* \in D$, there exists an open neighborhood in $\mathbb{R}^n$ entirely contained within D. The most elementary case arises when $D = \mathbb{R}^n$, which is often referred to as the fully unconstrained setting. From the perspective of local optimality analysis, however, it is sufficient to assume that x^* is an interior point of D, i.e., there exists $\varepsilon > 0$ such that (s.t.) $\mathbb{B}_\varepsilon(x^*) \subseteq D$. This condition is automatically satisfied whenever D is an open subset of $\mathbb{R}^n$.

2.1.1.1 First-Order Necessary Condition for Optimality

Suppose that f is a continuously differentiable function, i.e., $f \in C^1$, and let x^* be a local minimizer of f. Consider an arbitrary direction $d \in \mathbb{R}^n$. In the unconstrained setting, the point x^* is an interior point of the domain D, and hence there exists a scalar $\varepsilon > 0$ s.t.

$$x^* + \beta d \in D, \quad \forall \beta \in \mathbb{R} \text{ with } |\beta| < \varepsilon.$$

That is, sufficiently small perturbations of x^* along both d and $-d$ remain within the feasible set.

For a fixed direction $d \in \mathbb{R}^n$, consider the function obtained by restricting f along the line passing through x^* in the direction d. Specifically, define a scalar function $h : \mathbb{R} \to \mathbb{R}$ by

$$h(\beta) := f(x^* + \beta d),$$

where β belongs to an interval around 0 such that $x^* + \beta d \in D$.

Since x^* is a local minimizer of f, it follows that $\beta = 0$ is a local minimizer of h. This transformation is particularly useful, as it reduces the analysis of a multivariable function to that of a single-variable function, thereby allowing the application of standard tools from one-dimensional calculus. In particular, consider the first-order Taylor expansion of h about $\beta = 0$:

$$h(\beta) = h(0) + h'(0)\beta + o(\beta),$$

where $o(\beta)$ denotes higher-order terms satisfying

$$\lim_{\beta \to 0} \frac{o(\beta)}{\beta} = 0.$$

We claim that $h'(0) = 0$. To establish this, suppose, for the sake of contradiction, that $h'(0) \neq 0$. Then, by the definition of $o(\beta)$, there exists $\varepsilon > 0$ such that for all $\beta \neq 0$ with $|\beta| < \varepsilon$, we have $|o(\beta)| < |h'(0)\beta|$. Substituting into the Taylor expansion yields

$$h(\beta) - h(0) = h'(0) \ + o(\beta) < h'(0)\beta + |h'(0)\beta|.$$

Now, if β is chosen to have the opposite sign of $h'(0)$, the right-hand side becomes negative, implying

$$h(\beta) - h(0) < 0,$$

which contradicts the fact that $\beta = 0$ is a local minimum of h. Therefore, it must be that

$$h'(0) = 0.$$

We now express this result in terms of the original function f. By a direct application of the chain rule, the derivative of h can be written as

$$h'(\beta) = \nabla f(x^* + \beta d)^\top d,$$

where $\nabla f := \left(\frac{\partial f}{\partial x_1}, \ldots, \frac{\partial f}{\partial x_n}\right)^\top$ denotes the gradient of f, and $(\cdot)^\top d$ represents the standard inner product in $\mathbb{R}^n$. For brevity, we adopt the notation $f_{x_i} := \frac{\partial f}{\partial x_i}$ for partial derivatives.

Evaluating at $\beta = 0$, we obtain $h'(0) = \nabla f(x^*)^\top d$. Since it has already been established that $h'(0) = 0$, it follows that $\nabla f(x^*)^\top d = 0$. Because the direction $d \in \mathbb{R}^n$ is arbitrary, we conclude that

$$\nabla f(x^*) = 0,$$

which constitutes the first-order necessary condition (FONC) for optimality.

2.1.1.2 Second-Order Necessary Condition for Optimality

Under the stronger assumption that $f \in C^2$, we derive a second-order necessary condition (SONC) for optimality. Let x^* be a local minimizer and consider, for an arbitrary direction $d \in \mathbb{R}^n$, the function $h(\beta) = f(x^*+\beta d)$. Its second-order Taylor expansion about $\beta = 0$ is given by

$$h(\beta) = h(0) + h'(0)\beta + \frac{1}{2}h''(0)\beta^2 + o(\beta^2),$$

with

$$\lim_{\beta \to 0} \frac{o(\beta^2)}{\beta^2} = 0.$$

From the first-order condition, $h'(0) = 0$. Local minimality of h at $\beta = 0$ then implies

$$h''(0) \geq 0.$$

To express this in terms of f, differentiating $h'(\beta) = \nabla f(x^* + \beta d)^\top d$ yields

$$h''(\beta) = d^\top \nabla^2 f(x^* + \beta d)\, d,$$

and hence $h''(0) = d^\top \nabla^2 f(x^*)\, d$, where $\nabla^2 f(x^*)$ denotes the Hessian matrix of f at x^*.

Since d is arbitrary, it follows that $\nabla^2 f(x^*) \succeq 0$, i.e., the Hessian is positive semidefinite. This constitutes the SONC for optimality.

Remark 2.1 Similar to the FONC, the second-order condition is valid only in the unconstrained setting. However, it requires stronger smoothness assumptions, namely that $f \in C^2$, in contrast to the C^1 requirement for the first-order condition.

Moreover, unlike the first-order condition—which is satisfied at any extremum—the second-order condition provides a distinction between minima and maxima. In particular, at a local minimum the Hessian matrix is positive semidefinite, whereas at a local maximum it is negative semidefinite.

2.1.1.3 Second-Order Sufficient Condition for Optimality

A refinement of the SONC, combined with the first-order condition, yields a sufficient criterion for local optimality. Specifically, let $f \in C^2$ and let x^* be an interior point of its domain. If

$$\nabla f(x^*) = 0 \quad \text{and} \quad \nabla^2 f(x^*) \succ 0,$$

i.e., the Hessian at x^* is PD, then x^* is a *strict local minimizer* of f.

To justify this, consider $h(\beta) = f(x^* + \beta d)$ for an arbitrary direction $d \in \mathbb{R}^n$. Using the second-order expansion and the condition $\nabla f(x^*) = 0$, we obtain

$$f(x^* + \beta d) = f(x^*) + \frac{1}{2} d^\top \nabla^2 f(x^*) d\, \beta^2 + o(\beta^2).$$

Since $\nabla^2 f(x^*) \succ 0$, the quadratic term is strictly positive for all $d \neq 0$, and dominates the higher-order term $o(\beta^2)$ for sufficiently small α. Consequently, for all sufficiently small nonzero β, we have

$$f(x^* + \beta d) > f(x^*),$$

which establishes strict local minimality along any direction.

Finally, by restricting d to the unit sphere and invoking its compactness, one can ensure the existence of a uniform neighborhood around x^* in which the above inequality holds. Hence, x^* is a strict local minimum of f.

Definition 2.1 (Feasible Direction) Let $D \subseteq \mathbb{R}^n$ and let $x \in D$. A vector $d \in \mathbb{R}^n$ is called a *feasible direction* at x (with respect to D) if there exists a scalar $\varepsilon > 0$ s.t.

$$x + \alpha d \in D, \quad \forall \alpha \in [0, \varepsilon).$$

2.1.2 Convex Sets and Convex Functions

Convexity plays a central role in optimization theory, as it provides structure that ensures tractability and strong optimality properties. We begin by recalling the notion of convex sets (Fig. 2.1).

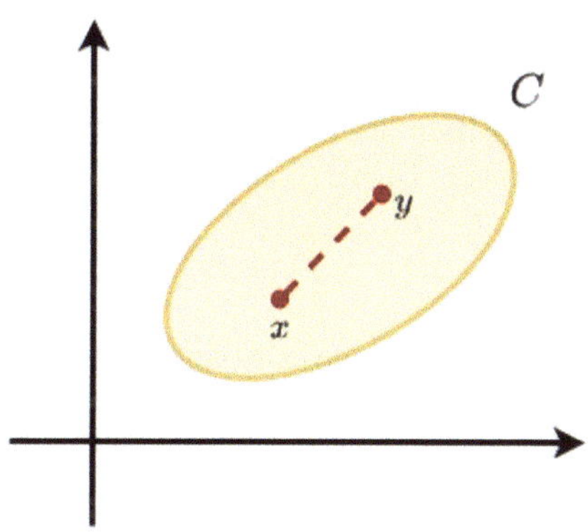

Fig. 2.1 Convex set

Fig. 2.2 Convex function

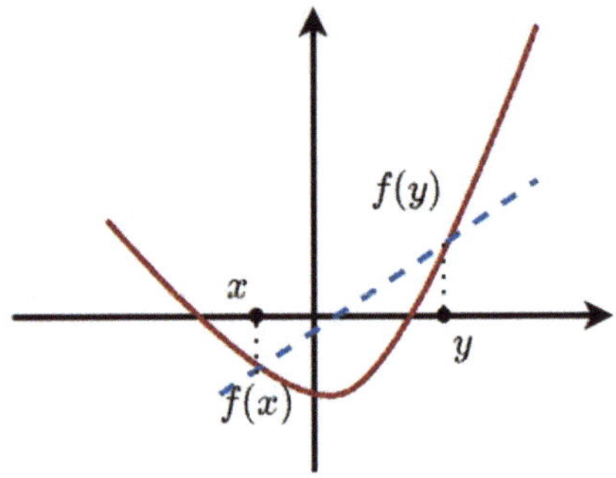

Definition 2.2 (Convex Set) A set $C \subseteq \mathbb{R}^n$ is said to be *convex* if for any $x, y \in C$ and any $\phi \in [0, 1]$, the point $\phi x + (1 - \phi)y \in C$.

The expression $\phi x + (1 - \phi)y$ is referred to as a *convex combination* of the points x and y. Thus, a set is convex if it contains the entire line segment joining any two of its points.

Definition 2.3 (Convex Function) Let $f : \mathbb{R}^n \to \mathbb{R}$ be defined on a convex set $C \subseteq \mathbb{R}^n$. The function f is said to be *convex* on C if for all $x, y \in C$ and all $\phi \in [0, 1]$,

$$f(\phi x + (1 - \phi)y) \leq \phi f(x) + (1 - \phi)f(y).$$

This definition states that the function value at a convex combination of two points does not exceed the corresponding convex combination of the function values, reflecting the intuitive notion that the graph of a convex function lies below the line segment joining any two points on it (Fig. 2.2).

Definition 2.4 (Strict Convexity) A function f is said to be *strictly convex* on C if for all $x \neq y$ in C and all $\theta \in (0, 1)$,

$$f(\phi x + (1 - \phi)y) < \phi f(x) + (1 - \phi)f(y).$$

Proposition 2.1 (First-Order Condition for Convexity) *Let $f \in C^1$ and $C \subseteq \mathbb{R}^n$ be convex. Then f is convex on C if and only if*

$$f(y) \geq f(x) + \nabla f(x)^\top (y - x), \quad \forall x, y \in C.$$

If $f \in C^2$, convexity can be characterized in terms of the Hessian.

Proposition 2.2 (Second-Order Condition for Convexity) *Let $f \in C^2$. Then f is convex on C if and only if*

$$\nabla^2 f(x) \succeq 0, \quad \forall x \in C,$$

i.e., the Hessian matrix is positive semidefinite for all $x \in C$.

Remark 2.2 Convex optimization problems possess a particularly favorable structure: every local minimizer is also a global minimizer. This property fundamentally distinguishes convex problems from general nonconvex optimization problems.

2.1.3 *Strong Convexity*

While convexity ensures that every local minimizer is global, stronger notions of curvature lead to improved uniqueness and convergence properties. One such notion is *strong convexity*.

Definition 2.5 (Strongly Convex Function) Let $C \subseteq \mathbb{R}^n$ be convex and let $f : C \to \mathbb{R}$. The function f is said to be *strongly convex* with parameter $\mu > 0$ on C if for all $x, y \in C$ and all $\phi \in [0, 1]$,

$$f(\phi x + (1-\phi)y) \leq \phi f(x) + (1-\phi)f(y) - \frac{\mu}{2}\phi(1-\phi)\|x - y\|^2.$$

Strong convexity strengthens the standard convexity inequality by introducing a quadratic term, which enforces a uniform curvature of the function.

For differentiable functions, strong convexity admits equivalent characterizations.

Proposition 2.3 (First-Order Characterization) *Let $f \in C^1$. Then f is strongly convex with parameter $\mu > 0$ if and only if*

$$f(y) \geq f(x) + \nabla f(x)^\top (y - x) + \frac{\mu}{2}\|y - x\|^2, \quad \forall x, y \in C.$$

Proposition 2.4 (Second-Order Characterization) *Let $f \in C^2$. Then f is strongly convex with parameter $\mu > 0$ if and only if*

$$\nabla^2 f(x) \succeq \mu I, \quad \forall x \in C,$$

i.e., the Hessian is uniformly PD.

Strong convexity has several important implications for optimization.

Proposition 2.5 (Uniqueness of Minimizer) *If f is strongly convex on C, then it admits a unique global minimizer in C.*

Proposition 2.6 (Quadratic Growth Condition) *If f is strongly convex with parameter $\mu > 0$ and x^* is its minimizer, then*

$$f(x) - f(x^*) \geq \frac{\mu}{2}\|x - x^*\|^2, \quad \forall x \in C.$$

2.1.4 Polyak–Lojasiewicz Inequality

Strong convexity is a powerful condition, but it can be restrictive in many applications. A weaker condition, which still guarantees favorable convergence properties, is the *Polyak–Lojasiewicz (PL) inequality*.

Definition 2.6 (Polyak–Lojasiewicz Inequality) Let $f : \mathbb{R}^n \to \mathbb{R}$ be a continuously differentiable function. We say that f satisfies the *Polyak–Lojasiewicz (PL) inequality* with parameter $\mu > 0$ if

$$\frac{1}{2}\|\nabla f(x)\|^2 \geq \mu\big(f(x) - f^*\big), \quad \forall x \in \mathbb{R}^n,$$

where $f^* := \inf_x f(x)$.

Proposition 2.7 *If f is strongly convex with parameter $\mu > 0$, then it satisfies the PL inequality with the same parameter. However, the converse is not true in general; there exist nonconvex functions that satisfy the PL inequality.*

2.1.5 Constrained Optimization

We now consider the case where the feasible set $D \subseteq \mathbb{R}^n$ is defined by a collection of equality constraints of the form

$$g_1(x) = g_2(x) = \cdots = g_m(x) = 0,$$

where each $g_i : \mathbb{R}^n \to \mathbb{R}$, $i = 1, \ldots, m$, is a C^1 function. The set D thus defines a constraint surface embedded in $\mathbb{R}^n$.

Assuming that the objective function $f : \mathbb{R}^n \to \mathbb{R}$ is also continuously differentiable, we are interested in characterizing its minima over the set D.

2.1.5.1 First-Order Necessary Condition

Let $x^* \in D$ be a local minimizer of f over D. We assume that x^* is a *regular point* of D, in the sense that the gradients $\nabla g_i(x^*)$, $i = 1, \ldots, m$, are linearly independent. This regularity condition excludes degenerate cases and ensures a well-defined local structure of the constraint set.

In contrast to the unconstrained case, where line segments through x^* were considered, we now restrict attention to curves lying entirely within D. Specifically, let $x(\beta) \in D$ be a continuously differentiable curve parameterized by $\beta \in \mathbb{R}$, such that $x(0) = x^*$.

For any such curve, define the scalar function $h(\beta) := f(x(\beta))$. This formulation generalizes the line-based construction used in the unconstrained setting. Since x^*

is a local minimizer over D, it follows that $\beta = 0$ is a local minimizer of h, and hence

$$h'(0) = 0.$$

Applying the chain rule, we obtain $h'(\beta) = \nabla f(x(\beta))^\top \dot{x}(\beta)$, and in particular, $h'(0) = \nabla f(x^*)^\top \dot{x}(0) = 0$. The vector $\dot{x}(0) \in \mathbb{R}^n$ plays a central role in this analysis. From the first-order Taylor expansion

$$x(\beta) = x^* + \dot{x}(0)\beta + o(\beta),$$

it follows that $\dot{x}(0)$ provides a linear approximation of the curve $x(\cdot)$ at x^*. Geometrically, it represents the instantaneous direction of motion along the curve at x^*, and is therefore referred to as a *tangent vector* to the set D at x^*. The collection of all such vectors forms the *tangent space*, denoted by $T_{x^*}D$.

To characterize $T_{x^*}D$ explicitly, recall that D is defined by the equality constraints

$$g_i(x) = 0, \quad i = 1, \ldots, m.$$

Since the curve $x(\beta)$ lies entirely in D, it satisfies $g_i(x(\beta)) = 0, \quad \forall \beta, \ i = 1, \ldots, m$. Differentiating with respect to β, we obtain

$$\frac{d}{d\beta} g_i(x(\beta)) = \nabla g_i(x(\beta))^\top \dot{x}(\beta) = 0.$$

Evaluating at $\beta = 0$, and using $x(0) = x^*$, yields

$$\nabla g_i(x^*)^\top \dot{x}(0) = 0, \quad i = 1, \ldots, m.$$

We have established that, for any C^1 curve $x(\cdot)$ in D with $x(0) = x^*$, its tangent vector $\dot{x}(0)$ satisfies $\nabla g_i(x^*)^\top \dot{x}(0) = 0, \quad i = 1, \ldots, m$. In fact, the converse also holds: any vector $d \in \mathbb{R}^n$ satisfying

$$\nabla g_i(x^*)^\top d = 0, \quad i = 1, \ldots, m,$$

can be realized as the tangent vector of some admissible curve in D passing through x^*. (This result relies on the regularity of x^* and is stated without proof.)

Thus, the tangent space $T_{x^*}D$ admits the characterization

$$T_{x^*}D = \left\{ d \in \mathbb{R}^n \;\middle|\; \nabla g_i(x^*)^\top d = 0, \ i = 1, \ldots, m \right\},$$

which is clearly a linear subspace of $\mathbb{R}^n$. Returning to the optimality condition, we have

$$\nabla f(x^*)^\top d = 0, \quad \forall d \in T_{x^*}D.$$

Using the above characterization of the tangent space, this can be equivalently expressed as

$$\nabla f(x^*)^\top d = 0 \quad \forall d \text{ such that } \nabla g_i(x^*)^\top d = 0,\ i = 1, \ldots, m.$$

This formulation, however, is implicit, as it involves all admissible directions d. It is therefore desirable to eliminate d and obtain a more explicit relationship between $\nabla f(x^*)$ and the constraint gradients $\nabla g_i(x^*)$, which leads naturally to the method of Lagrange multipliers. The gradient of f at x^* must lie in the linear span of the gradients of the constraint functions, i.e.,

$$\nabla f(x^*) \in \operatorname{span}\{\nabla g_i(x^*),\ i = 1, \ldots, m\}.$$

To see this, suppose otherwise. Then $\nabla f(x^*)$ would possess a nonzero component orthogonal to this span, implying the existence of a direction $d \neq 0$ such that

$$\nabla g_i(x^*)^\top d = 0, \quad i = 1, \ldots, m,$$

but $\nabla f(x^*)^\top d \neq 0$, contradicting the necessary optimality condition.

Geometrically, this indicates that $\nabla f(x^*)$ is orthogonal to the feasible set D at x^*. Consequently, there exist scalars $\lambda_1^*, \ldots, \lambda_m^* \in \mathbb{R}$ such that

$$\nabla f(x^*) + \sum_{i=1}^{m} \lambda_i^* \nabla g_i(x^*) = 0.$$

This relation represents the FONC for constrained optimality, with λ_i^* denoting the *Lagrange multipliers*.

Chapter 3
Mathematical Preliminaries

Abstract This chapter introduces standard notations, comparison functions, and key stability concepts, including Lyapunov, finite-time, fixed-time, and predefined-time stability. It further develops Lyapunov-based conditions for predefined-time convergence and discusses the existence and uniqueness of solutions. Additionally, input-to-state stability notions are introduced to analyze robustness under disturbances. These foundational tools provide the basis for the theoretical developments presented in the subsequent chapters.

3.1 Notations

Throughout this monograph, standard notation is adopted. The set of real numbers is denoted by $\mathbb{R}$. The sets of positive and non-negative real numbers are denoted by $\mathbb{R}_+ = \{x \in \mathbb{R} \mid x > 0\}$ and $\mathbb{R}_{\geq 0}$, respectively. Likewise, for a given constant $j \in \mathbb{R}$, the sets $\mathbb{R}_{>j}$ and $\mathbb{R}_{\geq j}$ represent the real numbers greater than j and greater than or equal to j, respectively.

The space $\mathbb{R}^n$ denotes the n-dimensional Euclidean space. For any $q \geq 1$, the ℓ_q-norm of a vector $z \in \mathbb{R}^n$ is defined as $\|z\|_q = \left(\sum_{i=1}^n |z_i|^q\right)^{\frac{1}{q}}$. When the norm is not explicitly specified, it is assumed to be the Euclidean norm, i.e., $\|z\| = \|z\|_2$. For matrices, the transpose of a matrix C is denoted by $C^\top$. For a vector $c = [c_1, c_2, \ldots, c_n] \in \mathbb{R}^n$, the notation $\text{diag}(c)$ denotes the diagonal matrix with diagonal entries c_i, for $i = 1, 2, \ldots, n$.

Furthermore, for a Lebesgue measurable function $\hat{l} : \mathbb{R}_{\geq 0} \to \mathbb{R}^m$, its norm over an interval $[t_1, t_2)$ is defined as $\|\hat{l}\|_{[t_1,t_2)} = \text{ess sup}_{t\in[t_1,t_2)} \|\hat{l}(t)\|$. In particular, the ∞-norm is given by $\|\hat{l}\|_\infty = \|\hat{l}\|_{[0,\infty)}$. A set of essentially bounded and measurable functions $\hat{l}$ with $\|\hat{l}\|_\infty < +\infty$ is represented as $\mathcal{L}_\infty$. The notation $f \in C^p(X, Y)$ denotes a function which is p times continuously differentiable, where $f : X \to Y$, $X \subseteq \mathbb{R}^n$, $Y \subseteq \mathbb{R}^m$. Consider a matrix E; we represent its null space and range space by $\mathcal{N}(E)$ and $\mathcal{R}(E)$, respectively. $\perp$ denotes orthogonality. For a given set S, the notation $\min\{S\}$ denotes its smallest element, while $\sup\{S\}$ represents its supremum,

S. Kamal, S. Pandey, *A Continuous-Time Perspective on Accelerated Optimization and Control*, Studies in Systems, Decision and Control 693,
https://doi.org/10.1007/978-3-032-31645-5_3

i.e., the least upper bound. An n-dimensional open ball of radius δ centered at $x_0 \in \mathbb{R}^n$ is defined as $B_\delta = \{x \in \mathbb{R}^n \mid \|x - x_0\| < \delta\}$. Furthermore, the absolute value function is denoted by $|\cdot|$. The transpose of a vector is denoted by $(\cdot)^\top$, and $\lceil x \rceil$ represents the ceiling function, i.e., the smallest integer greater than or equal to x. The space of bounded functions is denoted by ℓ_∞. Furthermore, $\min(\cdot, \cdot)$ denotes the minimum operator, which returns the smaller of its arguments at any given instant.

To effectively comprehend the developments presented in this work, the reader is expected to possess a fundamental understanding of the terminology and concepts commonly employed in the analysis of stability for continuous- and discrete-time dynamical systems. The definitions and notions introduced in this chapter play a crucial role throughout the subsequent sections of the monograph.

3.2 Comparison Functions

This section introduces a class of auxiliary functions that play a fundamental role in the characterization of stability properties of dynamical systems.

Definition 3.1 (Class $\mathcal{K}$ Function [1]) A function $\alpha : \mathbb{R}_{\geq 0} \to \mathbb{R}_{\geq 0}$ is said to belong to class $\mathcal{K}$ if it is continuous, strictly increasing, and satisfies $\alpha(0) = 0$. Furthermore, if $\alpha(r) \to \infty$ as $r \to \infty$, then α is said to be of class $\mathcal{K}_\infty$.

Definition 3.2 (Class $\mathcal{L}$ Function [1]) A function $\iota : \mathbb{R}_{\geq 0} \to \mathbb{R}_{\geq 0}$ is said to belong to class $\mathcal{L}$ if it is continuous, strictly decreasing, and satisfies $\lim_{r\to\infty} \iota(r) = 0$.

Definition 3.3 (Class $\mathcal{KL}$ Function [1]) A function $\beta : \mathbb{R}_{\geq 0} \times \mathbb{R}_{\geq 0} \to \mathbb{R}_{\geq 0}$ is said to belong to class $\mathcal{KL}$ if, for each fixed $t \geq 0$, the mapping $r \mapsto \beta(r, t)$ is of class $\mathcal{K}$, and for each fixed $r \geq 0$, the mapping $t \mapsto \beta(r, t)$ is of class $\mathcal{L}$.

Lemma 3.1 ([1]) *Let $\alpha_1, \alpha_2 : [0, a) \to \mathbb{R}_{\geq 0}$ be class $\mathcal{K}$ functions, and let $\alpha_3, \alpha_4 : \mathbb{R}_{\geq 0} \to \mathbb{R}_{\geq 0}$ be class $\mathcal{K}_\infty$ functions. Let β be a class $\mathcal{KL}$ function. Then, the following properties hold:*

- *The inverse function α_1^{-1} exists on $[0, \alpha_1(a))$ and belongs to class $\mathcal{K}$.*
- *The inverse function α_3^{-1} exists on $\mathbb{R}_{\geq 0}$ and belongs to class $\mathcal{K}_\infty$.*
- *The composition $\alpha_1 \circ \alpha_2$ is a class $\mathcal{K}$ function.*
- *The composition $\alpha_3 \circ \alpha_4$ is a class $\mathcal{K}_\infty$ function.*

Definition 3.4 (Generalized $\mathcal{K}$ Function [2]) A function $\alpha : [0, a) \to \mathbb{R}_{\geq 0}$ is called a generalized $\mathcal{K}$ function if it is continuous, nondecreasing, and satisfies $\alpha(0) = 0$. Moreover, it fulfills the following conditions: for any $r_1 > r_2$,

$$\begin{cases} \alpha(r_1) > \alpha(r_2), & \text{if } \alpha(r_1) > 0, \\ \alpha(r_1) = \alpha(r_2), & \text{if } \alpha(r_1) = 0. \end{cases}$$

Definition 3.5 (Generalized $\mathcal{KL}$ Function [2, 3]) A function $\Phi : \mathbb{R}_{\geq 0} \times \mathbb{R}_{\geq 0} \to \mathbb{R}_{\geq 0}$ is said to be of generalized $\mathcal{KL}$ type (denoted as $\mathcal{GKL}$) if, for every fixed $t \geq 0$, the mapping $r \mapsto \Phi(r, t)$ is a generalized $\mathcal{K}$ function, and for each fixed $r \geq 0$, the function $t \mapsto \Phi(r, t)$ is continuous and monotonically decreasing to zero as $t \to T$, where $T \leq \infty$. If the convergence time T can be specified a priori, then Φ is referred to as a predefined generalized $\mathcal{KL}$ (or $\mathcal{PGKL}$) function.

3.3 Stability Notions

In the present section, the notion of Predefined/Prescribed-time stability and the related previous results are discussed. Consider the following autonomous dynamical system:

$$\dot{\zeta} = f(t, \zeta), \qquad \zeta(t_0) = \zeta_0, \tag{3.1}$$

where $\zeta \in \mathbb{R}^n$ denotes the system state, $f : \mathbb{R}_{\geq 0} \times \mathbb{R}^n \to \mathbb{R}^n$ is continuous on an open neighborhood $D \subseteq \mathbb{R}^n$, and $f(t, 0) = 0$, i.e., $\zeta = 0$ is an equilibrium point of (3.1), t_0 is the initial time, and ζ_0 is the initial state. First, various notions of stability for autonomous systems are reviewed.

Definition 3.6 The origin of the system (3.1) is said to be:

- **Lyapunov stable [1]**, if for every $\epsilon > 0$, there exists $\delta(\epsilon) > 0$ such that if $\|\zeta(0)\| < \delta$, then $\|\zeta(t)\| < \epsilon$ for all $t \geq 0$.
- **Asymptotic stable [1]**, (respectively globally asymptotic stable) if it is stable, and there exists $c > 0$ such that for all $\|\zeta(0)\| < c$ (respectively, for all $\zeta(0) \in \mathbb{R}^n$), the solution of (3.1) satisfies $\lim_{t\to\infty} \zeta(t) = 0$.
- **Exponentially stable [1]**, (respectively globally exponentially stable), if there exist $c, b, \beta > 0$ such that $\|\zeta(t)\| \leq be^{-\beta t}\|\zeta(0)\|$, for all $\|\zeta(0)\| < c$ (respectively, for all $\zeta(0) \in \mathbb{R}^n$), and $t \geq 0$.
- **Finite-time stable (FTS) [4, 5]**, if it is stable and there exists an open neighborhood (N) of the origin such that for all $\zeta(0) \in N\backslash\{0\}$, $\lim_{t\to T} \zeta(t) = 0$, where $T = T(\zeta(0)) < \infty$. The origin is a globally finite-time stable if $N = \mathbb{R}^n$.
- **Fixed-time stable (FxTS) [6, 7]**, if it is globally FTS and the settling time function is uniformly bounded for all $\zeta(0) \in \mathbb{R}^n$, i.e., $\sup_{\zeta(0)\in\mathbb{R}^n} T(\zeta(0)) < \infty$.
- **Predefined-time stable (PTS) [8–10]**, if it is FxTS and there exists a $T_r > t_0$ which is chosen a priori for some value of system parameter $\eta > 1$ and is independent of the initial conditions, and the condition: $T_r \geq T_f$ can be established for some chosen $\eta > 1$, where T_f denotes the actual convergence time in which the system trajectories converge to the origin.

Remark 3.1 In contrast to fixed-time stability, where the UBST is finite but determined implicitly by system parameters, PTS ensures that the bound T_r can be explicitly assigned by design, independently of the initial condition.

3.4 Predefined-Time Stability

Consider the following nonautonomous differential equation

$$\dot{\zeta}(t) = \begin{cases} \frac{-\eta(e^{\zeta}-1)}{(e^{\zeta}(t_f-t))}, & \text{if } t_0 \leq t < t_f \\ 0, & \text{otherwise} \end{cases}$$

The solution to the above differential system is

$$\zeta(t) = \ln(C(t_f - t)^{\eta} + 1)$$

where integration constant $C = \frac{e^{\zeta(t_0)}-1}{(t_f-t_0)^{\eta}}$. Motivation of predefined-stability comes from the above non-autonomous differential equation, which makes both $\zeta(t)$ and $\dot{\zeta}(t) = 0$ tend to zero as $t \to t_f$.

Existence and Uniqueness of the Solution [11]

$$\dot{\zeta} = u_1; \quad \zeta \in \mathbb{R} \tag{3.2}$$

$$u_1 = \begin{cases} -\frac{\eta(e^{\zeta}-1)}{e^{\zeta}(t_f-t)} & \text{if } t_0 \leq t < t_f \\ 0, & \text{otherwise} \end{cases} \tag{3.3}$$

Let the initial condition be $\zeta(t_0) = \zeta_0$ and t_f is a prespecified time. Solution to (3.2) is: $\zeta = \ln\left(1 + C(t_f - t)^{\eta}\right)$, where, $C = \frac{e^{\zeta_0}-1}{(t_f-t_0)^{\eta}} \geq 0$, By substituting this solution in (3.3), we have,

$$u_1 = \begin{cases} \frac{-\eta\left(C(t_f-t)^{\eta-1}\right)}{(1+C(t_f-t)^{\eta})}, & \text{if } t_0 \leq t < t_f \\ 0, & \text{otherwise} \end{cases} \tag{3.4}$$

From (3.4), for $C \geq 1$, $|u_1| < \eta C, \forall t \geq t_0$ and for $C < 1$, $|u_1| \leq \eta, \forall t \geq t_0$, Hence, u_1 is continuous in ζ, $\forall$ fixed t, while measurable in t, $\forall$ fixed ζ and bounded by a quantity that is integrable over a finite-time interval. So, for $t < t_f$, all the conditions for the existence of solution are satisfied for (3.2) in Carathéodory sense, therefore, a solution is augmented till t_f. For $t > t_f$, a classical solution exists, and hence, we can say that the solution exists for $[t_0, \infty)$. The above analysis is the same as when solutions are analyzed in the sense of Filippov for differential equations with discontinuous right-hand side [12]. Hence, the existence and uniqueness of the solution are confirmed for the predefined-time dynamics.

Next, we discuss Lyapunov conditions adapted for predefined-time stability, adapted from [8].

Lemma 3.2 ([8]) *Consider the dynamics* (3.1)*, let $D \subseteq \mathbb{R}^n$ be a domain containing the equilibrium point $\zeta = 0$. Let $\gamma_1(\zeta), \gamma_2(\zeta)$ be continuous positive definite functions on D. Assume that there exists a real-valued continuously differentiable Lyapunov function candidate $V : \mathbb{R}_{\geq 0} \times D \to \mathbb{R}_{\geq 0}$ which satisfies:*

- $\gamma_1(\zeta) \leq V(t, \zeta) \leq \gamma_2(\zeta), \forall \zeta \in D \setminus \{0\}, \forall t$
- $V(t, 0) = 0, \forall t$
- *For* $V \neq 0, \dot{V} \leq \begin{cases} \frac{-\eta(e^V - 1)}{e^V(t_f - t)}, & \text{if } t_0 \leq t < t_f \\ 0, & \text{otherwise.} \end{cases}$

where $\eta > 1$. Then, the origin of (3.1) *is PTS, and $T_r = t_f - t_0$ (t_f can be chosen a priori).*

An alternate way to achieve the stability of the equilibrium point of a system in a prespecified-time is given in [13].

Lemma 3.3 ([13]) *Consider the system* (3.1)*. Let $D \subset \mathbb{R}^n$ contains $\zeta = 0$ and $\gamma_1(\zeta)$ and $\gamma_2(\zeta)$ be continuous positive definite functions. Assume that there exists a continuously differentiable scalar function $V : D \to \mathbb{R}_{\geq 0}$, and $\eta > 1$ such that*

- $\gamma_1(\zeta) \leq V(\zeta) \leq \gamma_2(\zeta), \forall \zeta \in D \setminus \{0\}, \forall t \in [t_0, \infty)$,
- $V(0) = 0, \forall t \in [t_0, \infty)$,
- $\frac{d}{dt} V(\zeta) \leq \begin{cases} \frac{-\eta V(\zeta)}{(t_f - t)}, & \text{if } t_0 \leq t < t_f \\ 0, & \text{otherwise.} \end{cases}$ *, for* $V \neq 0$,

Then the origin of system (3.1) *is prescribed-time stable and $T_r = t_f - t_0$ (t_f can be chosen a priori).*

3.5 Small-Disturbance Input to State Stability (SD-ISS)

Consider the following dynamics

$$\dot{\zeta}(t) = g(\zeta(t), d(t)), \tag{3.5}$$

where, $\zeta \in \mathcal{D} \subseteq \mathbb{R}^n$, $\mathcal{D}$ is an open subset of $\mathbb{R}^n$, and inputs $d(\cdot) : \mathbb{R}_{>0} \to \mathbb{R}^m$, $g : \mathcal{D} \times \mathbb{R}^m \to \mathbb{R}^n$.

Definition 3.7 ([14]) The function $V : \mathcal{D} \to \mathbb{R}$ is a size function if it is : (a) continuous, (b) $V(0) = 0$ and $V(\zeta) > 0$, $\forall \zeta \in \mathcal{D} \backslash \{0\}$, (c) coercive i.e. $\|\zeta_k\| \to \infty, V(\zeta_k) \to \infty$.

Definition 3.8 ([15]) Dynamics (3.5) is SD-ISS, if $\exists$ a size function $V(\cdot)$, a $d_0 \in \mathbb{R}$, a class $\mathcal{KL}$ function $\beta(\cdot,\cdot)$, and a class $\mathcal{K}$ function γ such that $\forall$ inputs d, locally bounded by d_0, $x(0) \in \mathcal{D}$, $\zeta(t)$ remains in $\mathcal{D}$ and follows

$$V(\zeta) \leq \beta(V(\zeta(0), t)) + \gamma(\|d\|_\infty), \ \forall t \geq 0. \tag{3.6}$$

Definition 3.9 ([14]) A function $V \in C^1(\mathcal{D}, \mathbb{R})$ is a SD-ISS Lyapunov function for dynamics (3.5) if (a) V is a size function, (b) $\exists$ a $\mathcal{K}$ class functions $\alpha_1(\cdot)$ and $\alpha_2(\cdot)$ such that if $\|\sigma\| \leq \alpha_1(V(\zeta))$, then

$$\dot{V}(\zeta, \sigma) \leq -\alpha_2(V(\zeta)). \tag{3.7}$$

Lemma 3.4 ([15]) *Dynamics* (3.5) *is a SD-ISS iff it exhibits a small disturbance ISS Lyapunov function.*

Definition 3.10 ([3]) The system (3.5) is PUBST-input to state stable (PUBST-ISS) if $\exists$ a $\mathcal{PGKL}$ class function $\tilde{\beta}$ with $\tilde{\beta}(r,t) = 0$, for $t \geq t_f$, where t_f is the predefined-time convergence, which is chosen in advance and a class $\mathcal{K}$ function $\tilde{\gamma}$ so that $\forall \zeta(t(0))$ and $d \in \mathcal{L}_\infty$, the solution $\zeta(t)$ exists and satisfies: $\|\zeta(t)\| \leq \tilde{\beta}(\|\zeta(t(0))\|, t_f - t) + \tilde{\gamma}(\|d\|_\infty), \ \forall t \geq 0$.

Definition 3.11 ([3]) A continuous function $V : \mathbb{R}_{\geq 0} \times \mathcal{D} \to \mathbb{R}$ is said as PUBST-ISS Lyapunov function for system (3.5) if it satisfies: (i) $\gamma_1(\|\zeta\|) \leq V(t,\zeta) \leq \gamma_2(\|\zeta\|), \ \forall \zeta \in \mathbb{R}^n$, where $\gamma_1, \gamma_2 \in \mathcal{K}_\infty$, (ii) $\dot{V}(t,\zeta) \leq -\tilde{\beta}(\|\zeta(t(0))\|, t_f - t) - \gamma_3(\|\zeta\|), \forall \|\zeta(t)\| \geq \gamma_4(\|d\|), \ \forall t \geq 0$, where $\gamma_3, \gamma_4 \in \mathcal{K}$ and $\tilde{\beta}$ is in $\mathcal{PGKL}$ class function and $\tilde{\beta}(\|\zeta\|, t_f - t) \mapsto \mu_1 \frac{e^V - 1}{e^V (t_f - t)}$, $\gamma_3(\|\zeta\|) \mapsto \mu_2 V^\rho$, where $\mu_1 > 1$, $\mu_2 > 0$ and $\rho > 0$.

Lemma 3.5 ([3]) *Consider the system* (3.5). *It is a small disturbance PUBST-ISS with input as d if it exhibits a PUBST-ISS Lyapunov function.*

References

1. Khalil, H.K.: Nonlinear Systems, 3rd edn. Prentice-Hall, Upper Saddle River (2002)
2. Hong, Y., Z.-P. Jiang, Feng, G.: Finite-time input-to-state stability and applications to finite-time control. IFAC Proc. Vol. **41**(2), 2466–2471 (2008)
3. Kamal, S., Pal, A.K., Singh, B., Bandyopadhyay, B., Yu, X.: Stabilization of continuous time uncertain systems with predefined upper bound of settling time. In: 2022 16th International Workshop on Variable Structure Systems (VSS), pp. 266–271. IEEE (2022)
4. Bhat, S.P., Bernstein, D.S.: Finite-time stability of continuous autonomous systems. SIAM J. Control Optim. **38**(3), 751–766 (2000)
5. Prasun, P., Kamal, S., Pandey, S., Bartoszewicz, A., Ghosh, S.: A minimum operator-based discrete-time sliding mode control. IEEE Trans. Autom. Control **69**(11), 7871–7876 (2024)
6. Polyakov, A.: Nonlinear feedback design for fixed-time stabilization of linear control systems. IEEE Trans. Autom. Control **57**(8), 2106–2110 (2011)

7. Pandey, S., Kamal, S., Osinenko, P., Parsegov, S., Singh, D.: Finite and fixed-time stabilization of discrete-time systems using passivity-based control. Int. J. Robust Nonlinear Control **34**(12), 8481–8494 (2024)
8. Pal, A.K., Kamal, S., Nagar, S.K., Bandyopadhyay, B., Fridman, L., Design of controllers with arbitrary convergence time. Automatica **112**, 108710 (2020)
9. Singh, B., Pal, A.K., Kamal, S., Dinh, T.N., Mazenc, F.: Vector control Lyapunov function based stabilization of nonlinear systems in predefined time. IEEE Trans. Autom. Control **68**, 4984 (2022)
10. Singh, B., Kamal, S., Patra, S., Ghosh, S.: Finite, fixed and prescribed-time stability and stabilization of nonlinear negative imaginary systems. Automatica **153**, 111003 (2023)
11. Pal, A.K., Kamal, S., Bandyopadhyay, B., Fridman, L.: On the predefined, prescribed and arbitrary time convergence. In: IECON 2022 – 48th Annual Conference of the IEEE Industrial Electronics Society, pp. 1–6 (2022)
12. Filippov, A.F.: Differential Equations with Discontinuous Righthand Sides: Control Systems, vol. 18. Springer Science & Business Media (2013)
13. Song, Y., Wang, Y., Holloway, J., Krstic, M.: Time-varying feedback for regulation of normal-form nonlinear systems in prescribed finite time. Automatica **83**, 243–251 (2017)
14. Sontag, E.D.: Remarks on input to state stability of perturbed gradient flows, motivated by model-free feedback control learning. Syst. Control Lett. **161**, 105138 (2022)
15. Sontag, E.D., Wang, Y.: On characterizations of the input-to-state stability property. Syst. Control Lett. **24**(5), 351–359 (1995)

Chapter 4
Accelerated Gradient Flows to Solve Convex Optimization Problems

Abstract This chapter develops novel predefined-time convergent gradient flow dynamics for solving convex optimization problems within a user-specified time bound. For unconstrained problems, modified gradient flow systems are proposed under the Polyak–Lojasiewicz condition, ensuring convergence independent of initial conditions and without complex parameter tuning. The framework is extended to constrained optimization with linear equality constraints using projected gradient and projected Newton flows, guaranteeing feasibility and optimality. Rigorous Lyapunov-based analysis establishes predefined-time stability, while robustness is addressed through input-to-state stability concepts. Numerical examples, including least squares and resource allocation problems, demonstrate the effectiveness and accelerated convergence of the proposed methods.

4.1 Introduction

This chapter first investigates unconstrained optimization problems through dynamical GF systems. For strongly convex objective functions, a GF with PTC is discussed, which is suitable for relaxing the objective function to only satisfy the P-L condition. In our schemes, the upper bound of convergence time can be chosen a priori, uniform with respect to initial conditions. Additionally, this approach does not involve complex parameter calculations. The convergence of proposed dynamics is rigorously proved by using the Lyapunov method. Further, in this chapter, we utilize the prospect of integrating the idea of a predefined-time stable gradient flow dynamics and projected gradient techniques to solve convex optimization problems with linear equality constraints (LEC). We propose a projected gradient flow algorithm with a predefined-time convergence property, where the UBST can be selected a priori. A perturbed projected gradient flow is also studied within the framework of ISS. Examples based on linear least squares and resource allocation problems (RAP) are solved using the proposed dynamics and simulation results illustrate the effectiveness of the proposed methods.

S. Kamal, S. Pandey, *A Continuous-Time Perspective on Accelerated Optimization and Control*, Studies in Systems, Decision and Control 693,
https://doi.org/10.1007/978-3-032-31645-5_4

The present chapter is structured as follows: Sect. 4.2 presents the problem statement. Section 4.3 introduces modified GF dynamics for solving unconstrained optimization problems. Section 4.5 focuses on solving convex optimization problems with LEC using predefined-time gradient flow dynamics. The efficacy of the methods proposed in Sects. 4.3 and 4.5 is further validated in Sects. 4.4 and 4.6, respectively, using illustrative examples like least squares.

4.2 Problem Statement

Consider the following unconstrained optimization problem

$$\min_{x \in \mathbb{R}^n} \mathcal{G}(x)$$

where $\mathcal{G} : D \to \mathbb{R}$ is a continuously differentiable function and $x \in \mathbb{R}^n$ is the decision variable. Suppose that $\mathcal{G}(x)$ satisfies that the gradient is locally Lipschitz continuous, and the optimal value of $\mathcal{G}(x)$ is attained, i.e., there exists x^* such that $\infty < \mathcal{G}^* = \mathcal{G}(x^*)$. Additionally, if $\mathcal{G}$ is a convex function, then $\|\nabla \mathcal{G}(x^*)\| = 0$ is both necessary and sufficient condition for the global minimum.

Lemma 4.1 (Lojasiewicz's Inequality [1]) *Let $\mathcal{G} : \mathbb{R}^n \to \mathbb{R}$ be a real analytic function on a neighbourhood of $y \in \mathbb{R}^n$. Then there exists a constant $d > 0$ and $\kappa \in [0, 1)$ s.t. $\|\nabla \mathcal{G}(x)\| \geq d|\mathcal{G}(x) - \mathcal{G}(y)|^{\kappa}$ holds in some neighbourhood of y.*

Assumption 2 (PL-Inequality (Gradient Dominance)) *The scalar function $\mathcal{G}$ is a continuously differentiable function. It has a unique minimizer $x = x^*$ and is gradient dominated, i.e., $\exists\, \beta > 0$, s.t.*

$$\frac{1}{2}\|\nabla \mathcal{G}(x)\|^2 \geq \beta(\mathcal{G}(x) - \mathcal{G}^{\star}) \quad \forall x \in \mathbb{R}^n \tag{4.1}$$

In what follows, we demonstrate that the solution to the proposed predefined-time convergent GF system corresponds to the solution of the previously stated optimization problem.

Remark 4.1 In this chapter, we adhere to the notions of stability aligned with the arbitrary-time stability as discussed in [2]. To ensure consistency with existing literature [3, 4], we adopt the term "Predefined-time stability" instead of arbitrary time stability.

4.3 Predefined-Time Convergent Gradient Flow Systems

In this section, we study the predefined-time gradient systems, whose convergence time can be explicitly given as a parameter in advance.

Consider the following non-autonomous GF system

$$\dot{x}(t) = \begin{cases} -\vartheta\varphi(t, x), & \text{for } t_0 \leq t < t_f, \\ -\vartheta\nabla\mathcal{G}(x(t))^\top, & \text{for } t \geq t_f, \end{cases} \tag{4.2}$$

where $x(t) \in \mathbb{R}^n$, $\vartheta \in \mathbb{R}$ and satisfies $\vartheta \geq 1$, $\nabla\mathcal{G}(x(t)) = (\nabla\mathcal{G}_{x1}(x), \cdots, \nabla\mathcal{G}_{xn}(x)) \in \mathbb{R}^n$, $\varphi(t, x) = (\varphi_1, \varphi_2, \ldots, \varphi_n)^\top$, $\varphi_i := \frac{(e^{\nabla\mathcal{G}_{x_i}(x)}-1)}{e^{\nabla\mathcal{G}_{x_i}(x)}(t_f-t)}$, if $t_0 \leq t < t_f$, for $1 \leq i \leq n$.

Definition 4.1 (Predefined-Time Stable Gradient Flow System [5–7]) The origin of the non-autonomous GF system (4.2) is called the predefined-time stable GF system if

- it is fixed-time stable,
- there exists an explicit scalar t_f, which is independent of any initial conditions and can be chosen a priori, and the condition $t_f > t_a$ follows for all $x(0) \in \mathbb{R}^n$, where t_a is the actual convergence time.

Theorem 4.1 ([5]) *Consider the GF system* (4.2) *and the objective function* $\mathcal{G}$ *satisfies Assumption 2. Assume a continuously differentiable proper loss function* $V : \mathbb{R}^n \to \mathbb{R}$ *for* (4.2)*, then the system* (4.2) *is a predefined-time stable GF system.*

Proof Consider the following proper loss function as

$$V(x) = \frac{1}{2}(\mathcal{G}(x) - \mathcal{G}^*)^2, \tag{4.3}$$

We begin the analysis for the time interval $t_0 \leq t < t_f$. The time derivative of the appropriate loss function is given by

$$\begin{aligned} \dot{V}(x) =&(\mathcal{G}(x) - \mathcal{G}^*)\nabla\mathcal{G}(x(t))\big(\dot{x}\big) \\ =&(\mathcal{G}(x) - \mathcal{G}^*)\nabla\mathcal{G}(x(t))\big(-\vartheta\varphi(t, x)\big) \\ \leq& -\vartheta(\mathcal{G}(x) - \mathcal{G}^*)\nabla\mathcal{G}_{x_i}(x)\varphi_i(t, x) \text{ for any } i \\ \leq& -\vartheta\sqrt{2}V^{\frac{1}{2}}\frac{|\nabla\mathcal{G}_{x_i}(x)|(e^{|\nabla\mathcal{G}_{x_i}(x)|} - 1)}{e^{|\nabla\mathcal{G}_{x_i}(x)|}(t_f - t)} \end{aligned} \tag{4.4}$$

Using Lojasiewicz's Inequality given in Lemma 4.1, note that there exists at least one i such that

$$|\nabla \mathcal{G}_{x_i}(x)| \geq \frac{d}{\sqrt{n}}|\mathcal{G}(x) - \mathcal{G}^*|^\kappa \tag{4.5}$$

since otherwise

$$\begin{aligned}\|\nabla \mathcal{G}(x)\| =& \sqrt{(\nabla \mathcal{G}_{x_1})^2 + (\nabla \mathcal{G}_{x_2})^2 + \cdots + (\nabla \mathcal{G}_{x_n})^2} \\ <& d|\mathcal{G}(x) - \mathcal{G}^*|^\kappa\end{aligned}$$

Then, Eqs. (4.4) and (4.5) gives

$$\begin{aligned}\dot{V}(x) \leq& -\vartheta\sqrt{2}V^{\frac{1}{2}}\frac{d}{\sqrt{n}}|\mathcal{G}(x) - \mathcal{G}^*|^\kappa \frac{(e^{\frac{d}{\sqrt{n}}|\mathcal{G}(x)-\mathcal{G}^*|^\kappa} - 1)}{e^{\frac{d}{\sqrt{n}}|\mathcal{G}(x)-\mathcal{G}^*|^\kappa}(t_f - t)} \\ \leq& -\lambda\frac{\alpha(V^\sigma(x))(e^{\alpha(V^\sigma(x))} - 1)}{e^{\alpha(V^\sigma(x))}(t_f - t)}\end{aligned}$$

where $d > 0, 0 < \kappa < 1, \lambda = 2^{\frac{1+\kappa}{2}}\theta\frac{d}{c}, \sigma = \frac{\kappa}{2}, \alpha \in \mathcal{K}_\infty$. That means the trajectories of the GF systems converge to the minima of $\mathcal{G}(x)$ within a priori chosen time.

Further, we carry out the analysis for the system (4.2) for $t \geq t_f$. Taking the time derivative of the considered loss function along the system trajectories (4.2) for $t \geq t_f$, we get

$$\begin{aligned}\dot{V}(x) =& (\mathcal{G}(x) - \mathcal{G}^*)\nabla \mathcal{G}(x(t))\big(-\vartheta\nabla \mathcal{G}(x(t))\big)^\top \\ =& -\vartheta(\mathcal{G}(x) - \mathcal{G}^*)\|\nabla \mathcal{G}(x)\|^2\end{aligned}$$

Using the relation $(\mathcal{G}(x) - \mathcal{G}^*) = \sqrt{2}V^{\frac{1}{2}}$ we can write,

$$\begin{aligned}\dot{V}(x) \leq& -\vartheta\sqrt{2}V^{\frac{1}{2}}\frac{1}{2}(\mathcal{G}(x) - \mathcal{G}^*)^2 \\ \leq& -\vartheta\sqrt{2}\alpha(V^{\sigma_1})\end{aligned}$$

where $\sigma_1 > 1, \alpha \in \mathcal{K}_\infty$. Hence, for $t \geq t_f$, the GF system (4.2) is asymptotically stable. □

4.4 Numerical Example: Least Squares

In this section, we explore a least squares problem, which involves finding the best possible fit for a given dataset.

4.4.1 *Least Squares*

Consider the following least squares problem

$$\min_{x\in\mathbb{R}^2} \|Cx - d\|^2 \tag{4.6}$$

where $x \in \mathbb{R}^2$, $C \in \mathbb{R}^{3\times 2}$, $d \in \mathbb{R}^3$ and $\mathcal{G}(x) := \|Cx - d\|^2$. Matrix C is given by $C = \begin{bmatrix} 1 & 1 \\ 1 & 2 \\ 1 & 3 \end{bmatrix}$ and vector d is given by $d = \begin{bmatrix} 1 \\ 2 \\ 2 \end{bmatrix}$. Find the best possible solution by minimizing the objective function (4.6).

Note that the objective function (4.6) is a strongly convex problem, and hence the obtained solution will be valid globally.

While simulating for the predefined-time convergent GF scheme, the initial conditions were taken as $x(0) = [0 \;\; 0]^\top$. Figure 4.1 shows the evolution of the states, showing that the states reach optimal points within a predefined time $t_f = 0.05$ s which is chosen in advance, and comparison with gradient flow dynamics. In Fig. 4.1 also presents the corresponding curve of the gradient norm, clearly illustrating that the gradient becomes zero within the predefined time $t_f = 0.05$ s and the proposed dynamics is much more accurate than gradient flow dynamics. Figure 4.2 shows the contour plot for the proposed and standard GF dynamics.

4.5 Convex Optimization Problem with Linear Equality Constraints

Consider the following optimization problem

$$\begin{aligned} &\min_{x\in\mathbb{R}^n} \mathcal{F}(x) \\ &\text{subject to } C^\top x = d, \end{aligned} \tag{4.7}$$

where, $x \in \mathbb{R}^n$, $\mathcal{F} : \mathbb{R}^n \to \mathbb{R}$ is an objective function, $C \in \mathbb{R}^{n\times m}$, $d \in \mathbb{R}^m$. Let $\mathbb{X}_{feas} := \{x \in \mathbb{R}^n : C^\top x = d\}$ is the set of feasible points. The optimum of (4.7) is

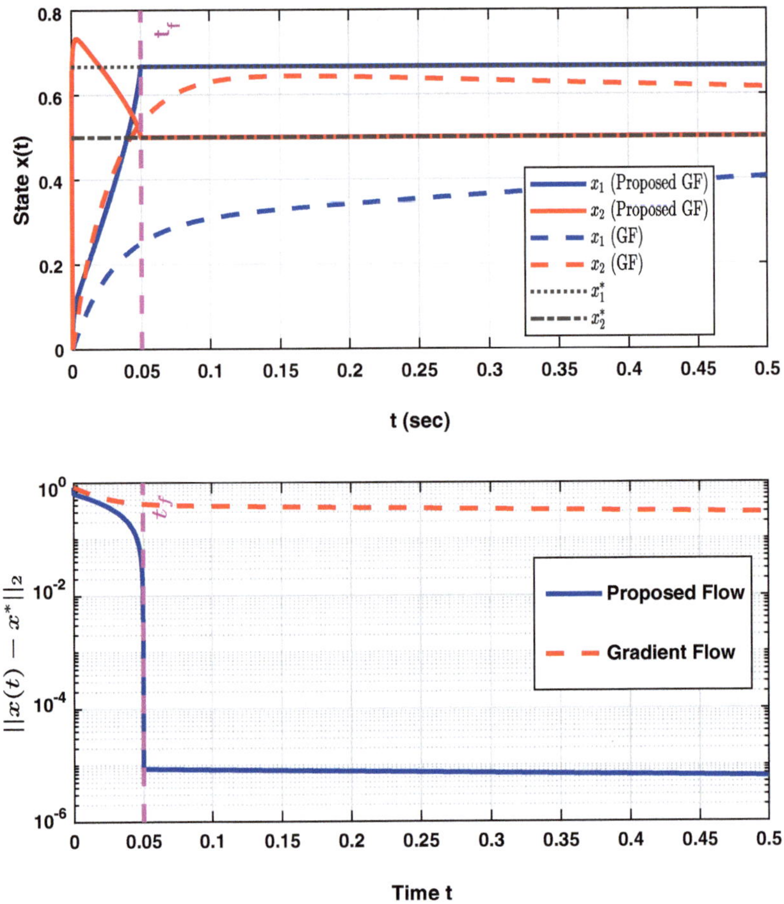

Fig. 4.1 Comparison of proposed flow with gradient flow showing convergence of states to the optimal points within $t_f = 0.05$ s and evolution of norm of the gradient

$\mathcal{F}^* = \inf_{x \in \mathbb{R}^n} \{\mathcal{F}(x) | C^\top x = d\}$. If $\mathcal{F}(x^*) = \mathcal{F}^*$ and $x^* \in \mathbb{X}_{feas}$, then we say x^* is an optimal point.

Assumption 3 *$\mathcal{F} \in C^2(\mathcal{D}, \mathbb{R})$ and $\mathcal{F}$ is convex.*

Assumption 4 *The columns of C are linearly independent.*

Lemma 4.2 ([8]) *Suppose Assumption 3 and 4 holds then x^* is the optimal solution to the* (4.7) *iff,*

$$
\begin{aligned}
C^\top x^* &= d \\
\nabla \mathcal{F}(x^*) &\in \mathcal{R}(C).
\end{aligned} \tag{4.8}
$$

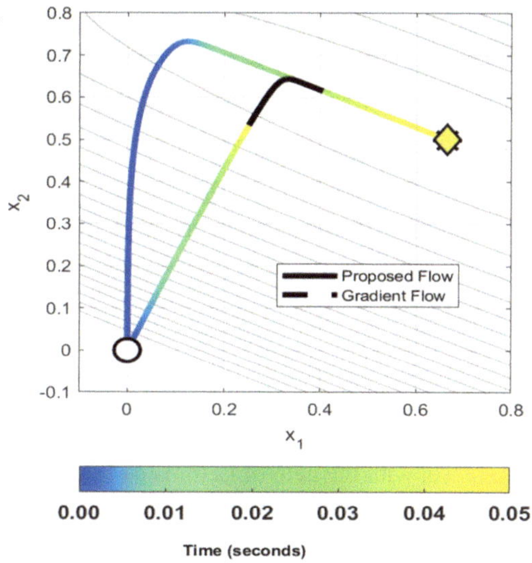

Fig. 4.2 Contour plot

4.5.1 Predefined-Time Convergent Projected Gradient Flow

In this subsection, we propose a dynamics that is capable of finding the optimal solution of (4.7) in a priori chosen time. The equilibrium point of the proposed dynamics coincides with the optimal point x^* of (4.7). We propose the following non-autonomous predefined-time convergent Projected Gradient Flow (PGF) dynamics:

$$\dot{x}(t) = \begin{cases} -\theta P^\top \mathrm{Sign}(\wp(x))\varphi(t, \wp(x)), & \text{for } t \in [t_0, t_f), \\ -\theta \wp(x), & \text{for } t \geq t_f, \end{cases} \tag{4.9}$$

where $x(t) \in \mathbb{R}^n$, $\theta \in \mathbb{R}$, $\theta > 1$, $\wp(x) = P\nabla\mathcal{F}(x)$, $\wp = (\wp_1, \ldots, \wp_n) \in \mathbb{R}^n$, $P \in \mathbb{R}^{n\times n}$ is projection matrix, $\varphi(t, \wp) = (\varphi_1, \varphi_2, \ldots, \varphi_n)^\top$, $\varphi_i(t, x) := \frac{(e^{|\wp_i(x)|}-1)}{e^{|\wp_i(x)|}(t_f - t)}$, for $1 \leq i \leq n$ and $\mathrm{Sign}(\wp) := \mathrm{diag}(\mathrm{sign}(\wp_1), \ldots, \mathrm{sign}(\wp_n)) \in \mathbb{R}^{n\times n}$.

Remark 4.2 The projection matrix (P) is structured to follow $\mathcal{N}(P) = \mathcal{R}(C)$.

Proposition 4.1 ([6]) *Assume that $x(t_0) \in \mathbb{X}_{feas}$, and Assumption 3, 4, and 4.2 holds. Then, $\mathbb{X}_{feas}$ is positively invariant under the dynamics* (4.9) *and the unique equilibrium point of* (4.9) *is the optimal solution of* (4.7).

Proof Since for a matrix S, $\mathcal{R}(S) \perp \mathcal{N}(S^\top)$ (see [9]), we get that $\mathcal{R}(P^\top) = \mathcal{N}(C^\top)$, which follows $C^\top P^\top = \mathbf{0}$. We consider the following Filippov time-varying set-valued map of the vector field in (4.9) for time interval $[t_0, t_f)$

$$\mathfrak{K}(\dot{x}) = -\theta P^\top l\varphi(t, \wp) \tag{4.10}$$

where $P \in \mathbb{R}^{n\times n}$, $\varphi \in \mathbb{R}^n$, $l := \text{diag}(l_1, \cdots, l_n) \in \mathbb{R}^{n\times n}$ and $l_i = \begin{cases} \{-1\}, & \text{sign}(\wp_i) < 0 \\ [-1, 1], & \text{sign}(\wp_i) = 0 \\ \{1\}, & \text{sign}(\wp_i) > 0 \end{cases}$. By carrying out multiplication of $C^\top$ on both sides of (4.10), we get $C^\top \mathfrak{K}(\dot{x}) = \mathbf{0}$. Hence, $\mathbb{X}_{feas}$ is positively invariant.

Furthermore, we demonstrate that the origin of (4.9) is the optimal solution of (4.7). The origin of (4.9) is calculated by equating $\dot{x} = 0$. For the interval $[t_0, t_f)$,

$$P^\top \text{Sign}(\wp)\varphi(t, \wp) = \mathbf{0} \tag{4.11}$$

By performing multiplication on both sides of (4.11) from the left by $(\nabla\mathcal{F})^\top$, we get $P\nabla\mathcal{F} = \mathbf{0}$. Similarly for $t > t_f$ also we get $P\nabla\mathcal{F} = \mathbf{0}$ So, $\nabla\mathcal{F}(x) \in \mathcal{N}(P) = \mathcal{R}(C)$.

Hence, with the help of Lemma 4.2 and positive invariance of $\mathbb{X}_{feas}$, we can say that the equilibrium point of (4.9) is the optimal point of (4.7). □

4.5.2 *Convergence Analysis*

In the present section, we perform predefined-time convergence analysis of (4.9) using Lyapunov's direct method.

Theorem 4.2 ([6]) *For any $t_f > 0$, consider the dynamics* (4.9)*. Let Assumptions 3, 4, and 4.2 are satisfied and $x(0) \in \mathbb{X}_{feas}$. Assume that $\exists$ a $\mathcal{V} \in C^1(\mathbb{R}^n, \mathbb{R})$ that adheres to $\psi_1(x) \leq \mathcal{V}(x) \leq \psi_2(x)$, $\forall x \in \mathbb{R}^n\backslash\{0\}$ where ψ_1 and ψ_2 are continuous positive definite functions, $\mathcal{V}(0) = 0$ and for $\mathcal{V} \neq 0$, $\dot{\mathcal{V}} \leq \frac{-\theta(e^{\mathcal{V}}-1)}{e^{\mathcal{V}}(t_f-t)}$, $\forall t \in [t, t_f), \theta > 1$ and $\dot{\mathcal{V}} \leq 0, \forall t \geq t_f$. Then, the equilibrium point of the projected GF system* (4.9) *is predefined-time stable.*

Proof Consider a candidate Lyapunov function as $\mathcal{V}(x) = (\mathcal{F} - \mathcal{F}^*)$, and when $\mathcal{V}(x) = 0$, $\mathcal{F}(x) = \mathcal{F}^*$ and optimal solution is reached. First, we perform the analysis for $t_0 \leq t < t_f$. By calculating the time derivative of considered Lyapunov function $\mathcal{V}(x)$ along the dynamics of (4.9), we get

$$\begin{aligned}\dot{\mathcal{V}} &= (\nabla\mathcal{F})^\top \dot{x} \\ &= -\theta(\nabla\mathcal{F})^\top P^\top \text{Sign}(\wp)\varphi(t, \wp) \\ &= -\theta\wp^\top \text{Sign}(\wp)\varphi(t, \wp) \\ &= -\theta|\wp|^\top \varphi(t, \wp)\end{aligned}$$

where $|\wp| = (|\wp_1|, \ldots, |\wp_n|) \in \mathbb{R}^n$. Next, for any $i \in (1, .., n)$

$$\dot{\mathcal{V}} \leq -\theta|\wp_i|\varphi_i \tag{4.12}$$

As $|\wp_i| \geq p_i |\nabla_i \mathcal{F}|$, where, $p_i > 0$. Next, from PL inequality in Lemma 4.1, there exists i, for which $|\nabla_i \mathcal{F}| \geq \sqrt{\frac{\alpha}{n}} (\mathcal{F} - \mathcal{F}^*)^{\frac{1}{2}}$. Therefore, we can write

$$\dot{\mathcal{V}} \leq -\theta\delta\sqrt{\mathcal{V}} \frac{(e^{\delta\sqrt{\mathcal{V}}} - 1)}{e^{\delta\sqrt{\mathcal{V}}}(t_f - t)},$$

where, $\delta := p_i \sqrt{\frac{\alpha}{n}}$. Let $\hat{\mathcal{V}} = \delta\sqrt{\mathcal{V}}$. Then $\dot{\hat{\mathcal{V}}} = \frac{\delta^2}{2} \frac{\dot{\mathcal{V}}}{\hat{\mathcal{V}}}$. Therefore,

$$\dot{\hat{\mathcal{V}}} \leq -\hat{\theta} \frac{(e^{\hat{\mathcal{V}}} - 1)}{e^{\hat{\mathcal{V}}}(t_f - t)}, \tag{4.13}$$

where $\hat{\theta} = \frac{\theta\delta^2}{2} > 1$. Hence, according to Lemma 3.2 inequality (4.13) shows that the equilibrium point of (4.9) is predefined-time stable. Next, we perform the analysis for $t \geq t_f$. By computing the time derivative of $\mathcal{V}(x)$ along the dynamics of (4.9) for $t \in [t_f, \infty)$, we get

$$\dot{\mathcal{V}} = -\theta (\nabla \mathcal{F})^\top \wp(x) \tag{4.14}$$

Since, $P^\top P = P$, it follows that

$$\dot{\mathcal{V}} \leq -\theta \wp^\top \wp \tag{4.15}$$

which represents that the origin of (4.9) is asymptotically stable for $t \geq t_f$. Hence, the trajectory of (4.9) will remain at its equilibrium point. Hence, the equilibrium point of PGF system is predefined-time stable. □

Remark 4.3 The selection of P may not be unique. One of the choice is $P = I_n - A(A^\top A)^{-1} A^\top$ which satisfies Assumption 4.2 (see [10]).

Computational Cost We can discuss the convergence rate of Euler discretization of (4.9) to calculate complexity, considering the assumption of Lipschitz smoothness. Let us consider the Euler discretized version of the proposed PUBST-based PGF dynamics. To simplify the computation of complexity, we can write $\varphi(t, x) := \frac{P\nabla\mathcal{F}}{t_f - t}$. Let $k_f = t_f/h$, then the forward Euler discretized version of the proposed PUBST-based PGF dynamics, taking h as a practical step size, can be written as:

$$x(k+1) = \begin{cases} x(k) - \frac{h\theta P^\top P\nabla\mathcal{F}}{k_f - k}, & for\ 0 \leq k < k_f, \\ x(k) - h\theta P \nabla\mathcal{F}, & for\ k \geq k_f \end{cases} \tag{4.16}$$

We need to introduce an extra assumption of L-Lipschitz smooth, as given below:

Assumption (Lipschitz Smoothness of Order q) We assume the function $\mathcal{F}$ is L-Lipschitz smooth of order $q \in (1, 2]$, i.e., for any $z, y \in \mathbb{R}^n$,

$$\|\nabla\mathcal{F}(y) - \nabla\mathcal{F}(x)\| \leq L\|y - x\|^{q-1}$$

The smoothness assumption is a prevalent setting in optimization algorithm literature. This assumption is also called (L, q) Holder continuity, it will lead to the following property:

$$\mathcal{F}(y) \leq \mathcal{F}(x) + \langle\nabla\mathcal{F}(x), y - x\rangle + \frac{L}{q}\|y - x\|^q,$$

When $q = 2$, the function will be Lipschitz smooth. Considering the following assumptions, for our case, we can calculate iteration complexity as follows:

$$\mathcal{F}(x(k+1)) \leq \mathcal{F}(x(k)) + \langle\nabla\mathcal{F}(x(k)), x(k+1) - x(k)\rangle + \frac{L}{q}\|x(k+1) - x(k)\|^q,$$

Using algorithm (4.16), for $0 \leq k < k_f$, considering the Lipschitz continuity of the gradient and $q = 2$,

$$\mathcal{F}(x(k+1)) \leq \mathcal{F}(x(k)) - \frac{h\theta}{k_f - k}\|P\nabla\mathcal{F}\|^2 + \frac{Lh^2\theta^2}{2(k_f - k)^2}\|P\nabla\mathcal{F}\|^2,$$

Let $\eta_k := \frac{\theta}{k_f - k}$, then we can write $\mathcal{F}(x(k+1)) \leq \mathcal{F}(x(k)) - \left(\eta_k - \frac{Lh\eta_k^2}{2}\right) h$ $\|P\nabla\mathcal{F}\|^2$,. Choose a step size such that $h < \frac{2k_f}{L\theta}$, and Let $\|P\nabla\mathcal{F}\|^2 > p_m\|\nabla\mathcal{F}\|^2$, where, $p_m > 0$. Hence, we can further write $\mathcal{F}(x(k+1)) \leq \mathcal{F}(x(k)) - \left(\eta_k - \frac{Lh\eta_k^2}{2}\right) hp_m\|\nabla\mathcal{F}\|^2$,. Let us define $\sigma_k := \left(\eta_k - \frac{Lh\eta_k^2}{2}\right)$ and using assumption of PL inequality: $\|\nabla\mathcal{F}\|^2 \geq \alpha(\mathcal{F}(x) - \mathcal{F}(x^*))$, then further

$$\mathcal{F}(x(k+1)) \leq \mathcal{F}(x(k)) - \sigma_k hp_m\alpha(\mathcal{F}(x(k)) - \mathcal{F}(x^*)),$$

Subtracting $\mathcal{F}(x^*)$ on both the sides, we get

$$\mathcal{F}(x(k+1)) - \mathcal{F}(x^*) \leq (1 - \sigma_k hp_m\alpha)(\mathcal{F}(x(k)) - \mathcal{F}(x^*)),$$

The for k_f iterations with fixed step size h,

$$\mathcal{F}(x(k_f)) - \mathcal{F}^* \leq (1 - \sigma_k hp_m\alpha)^{k_f}(\mathcal{F}(x(0)) - \mathcal{F}^*).$$

Using inequality $e^{-x} \geq 1 - x$, we can get that the corresponding iteration complexity $O\left((\sigma_k h p_m \alpha)^{-1} \ln \frac{1}{\epsilon}\right)$ for ϵ closeness of $x(k)$ to x^*. Similarly, if we run the algorithm for K iterations, then for $k_f \leq k \leq K$, the corresponding iteration complexity is $O\left((\sigma h p_m \alpha)^{-1} \ln \frac{1}{\epsilon}\right)$, where $\sigma := \left(1 - \frac{L}{2} h\theta\right)$.

4.5.3 Predefined-Time Convergent Projected Newton Flow

We propose the following non-autonomous, Predefined-Time Convergent Projected Newton Flow (PTPNF) dynamics:

$$\dot{x}(t) = \begin{cases} -\theta(\nabla^2\mathcal{F})^{-1} P^\top \mathrm{Sign}(\wp(x))\varphi(t, \wp(x)), & \text{for } t \in [t_0, t_f), \\ -\theta(\nabla^2\mathcal{F})^{-1}\wp(x), & \text{for } t \in [t_f, \infty), \end{cases} \tag{4.17}$$

where $x(t) \in \mathbb{R}^n$, $\theta > 1$, $\wp(x) = P\nabla\mathcal{F}(x)$, $\nabla^2\mathcal{F} : \mathbb{R}^n \to \mathbb{R}^{n\times n}$ denotes Hessian of the function $\mathcal{F}$, $\wp = (\wp_1, \ldots, \wp_n) \in \mathbb{R}^n$, $P \in \mathbb{R}^{n\times n}$ is projection matrix, $\varphi(t, \wp) = (\varphi_1, \varphi_2, \ldots, \varphi_n)^\top$, $\varphi_i(t, x) := \frac{(e^{|\wp_i(x)|}-1)}{e^{|\wp_i(x)|}(t_f - t)}$, for $1 \leq i \leq n$ and $\mathrm{Sign}(\wp) := \mathrm{diag}(\mathrm{sign}(\wp_1), \ldots, \mathrm{sign}(\wp_n)) \in \mathbb{R}^{n\times n}$.

Theorem 4.3 ([6]) *For any $t_f > 0$, consider the dynamics* (4.17)*. Let Assumptions 3, 4, and 4.2 are satisfied and $x(0) \in \mathbb{X}_{feas}$. Assume that $\exists$ a function $\mathcal{V} \in C^1(\mathbb{R}^n, \mathbb{R})$ which follows $\psi_1(x) \leq \mathcal{V}(x) \leq \psi_2(x)$, $\forall x \in \mathbb{R}^n\backslash\{0\}$ where ψ_1 and ψ_2 are continuous positive definite functions, $\mathcal{V}(0) = 0$ and for $\mathcal{V} \neq 0$, $\dot{\mathcal{V}} \leq \frac{-\theta(e^{\mathcal{V}}-1)}{e^{\mathcal{V}}(t_f - t)}$, $\forall t \in [t, t_f), \theta > 1$ and $\dot{\mathcal{V}} \leq 0, \forall t \geq t_f$. Then, the equilibrium point of the PTPNF system* (4.17) *is predefined-time stable.*

Proof Consider the candidate Lyapunov function as $\mathcal{V}(x) = \frac{1}{2}\|P\nabla\mathcal{F}\|^2$. Using Assumptions 3, 4, and 4.2, we can say that if $\mathcal{V}(x) = 0$, the conditions in Lemma 4.2 are satisfied and optimal solution is reached. First, we perform the analysis for $t \in [t_0, t_f)$. By calculating the time derivative of $\mathcal{V}(x)$ along the dynamics of (4.17), we get

$$\begin{aligned} \dot{\mathcal{V}} &= (P\nabla\mathcal{F})^\top P\nabla^2\mathcal{F}\dot{x} \\ &= -\theta\nabla\mathcal{F}^\top P^\top P P^\top \mathrm{Sign}(\wp(x))\varphi(t, \wp(x)) \end{aligned}$$

Note that the Projection matrix has the properties $P^\top = P,\ P^2 = P$. Next,

$$\dot{\mathcal{V}} = -\theta|\wp|^\top \varphi(t, \wp(x))$$

where $|\wp| = (|\wp_1|, \ldots, |\wp_n|) \in \mathbb{R}^n$. Next, for some $i \in (1, 2, .., n)$

$$\begin{aligned}\dot{\mathcal{V}} &\leq -\theta|\wp_i|\varphi_i \\ &\leq -\theta\delta_1\sqrt{\mathcal{V}}\frac{(e^{\delta_1\sqrt{\mathcal{V}}}-1)}{e^{\delta_1\sqrt{\mathcal{V}}}(t_f-t)},\end{aligned} \tag{4.18}$$

where, $\delta_1 = \frac{2}{\sqrt{n}}$. Let $\hat{\mathcal{V}} = \delta_1\sqrt{\mathcal{V}}$. Then $\dot{\hat{\mathcal{V}}} = \frac{\delta_1^2}{2}\frac{\dot{\mathcal{V}}}{\hat{\mathcal{V}}}$. Therefore,

$$\dot{\hat{\mathcal{V}}} \leq -\hat{\theta}\frac{(e^{\hat{\mathcal{V}}}-1)}{e^{\hat{\mathcal{V}}}(t_f-t)}, \tag{4.19}$$

where $\hat{\theta} = \frac{\theta\delta_1^2}{2} > 1$. Hence, according to Lemma 3.2, inequality (4.19) shows that the equilibrium point of (4.17) is predefined-time stable. Next, we perform the analysis for $t \geq t_f$. By computing the time derivative of $\mathcal{V}(x)$ along the dynamics of (4.17) for $t \geq t_f$, we get

$$\begin{aligned}\dot{\mathcal{V}} &= -\theta\nabla\mathcal{F}^\top P^\top\wp(x) \\ &= -\theta\mathcal{V},\end{aligned}$$

which represents that the origin of (4.17) is asymptotically stable for $t \geq t_f$. Hence, the trajectory of (4.17) will remain at its equilibrium point. Hence, the equilibrium point of the PTPNF system is predefined-time stable. □

4.5.4 Robustness Analysis of Perturbed Projected Gradient Flow

In this subsection, we discuss the effect of perturbations on PGF employing the formalism of Input to State Stability (ISS).

4.5.4.1 Proper Loss Functions (PLFs)

Definition 4.2 ((See [11])) A function $\mathcal{F} \in C^1(\mathbb{Y}, \mathbb{R})$ with locally Lipschitz gradient $\nabla\mathcal{F}$ is PLF with respect to $(\mathbb{Y}, \mathcal{M})$ if $\mathcal{F}(x) = \mathcal{F}^*$ for $x \in \mathcal{M}$ and satisfies: (i) $\mathcal{F} - \mathcal{F}^*$ is a size function for $(\mathbb{Y}, \mathcal{M})$, (ii) $|P\nabla\mathcal{F}(x)|$ is a size function for $(\mathbb{Y}, \mathcal{M})$.

Lemma 4.3 (See [11]) *Given a $\mathcal{F} \in C^1(\mathbb{Y}, \mathbb{R})$ with a locally Lipschitz gradient $\nabla\mathcal{F}$ such that $\mathcal{F}(x) - \mathcal{F}^*$ is a size function for $(\mathbb{Y}, \mathcal{M})$. Then, the following pair of characteristics are equivalent:*

- *$\mathcal{F}$ is a PLF,*
- $\alpha(\zeta(x)) \leq \|P\nabla\mathcal{F}(x)\|^2$, $\forall x \in \mathbb{Y}$, *where* $\alpha \in \mathcal{K}_\infty$ *and* ζ *denotes size function.*

Consider the following perturbed projected GF system:

$$\dot{x}(t) = \begin{cases} -\theta_1 P^\top \text{Sign}(\wp(x))\varphi(t, \wp(x)) - \theta_1 \nabla\mathcal{F} + \Delta x(t), & \text{for } t_0 \leq t < t_f, \\ -\theta_1 \wp(x) - \theta_1 \nabla\mathcal{F} + \Delta x(t), & \text{for } t \in [t_f, \infty), \end{cases} \tag{4.20}$$

where $\Delta x \in \mathbb{R}^m$ represents disturbance inputs, $x(t) \in \mathbb{R}^n$, $\theta_1 \in \mathbb{R}$, $\theta_1 > 1$, $\wp(x) = P\nabla\mathcal{F}(x)$, $\wp = (\wp_1, \ldots, \wp_n) \in \mathbb{R}^n$, $P \in \mathbb{R}^{n\times n}$ is projection matrix, $\varphi(t, \wp) = (\varphi_1, \ldots, \varphi_n)^\top$, $\varphi_i(t, x) := \frac{(e^{|\wp_i(x)|}-1)}{e^{|\wp_i(x)|}(t_f - t)}$, for $1 \leq i \leq n$ and $\text{Sign}(\wp) := \text{diag}(\text{sign}(\wp_1), \ldots, \text{sign}(\wp_n)) \in \mathbb{R}^{n\times n}$, and $\nabla\mathcal{F} \in \mathbb{R}^n$ as gradient vector.

Theorem 4.4 ([6]) *Let $\mathcal{F}$ be a proper loss function. Then, system* (4.20) *is a small disturbance Predefined Upper Bound of Settling Time Based -Input to state Stable (PUBST-ISS).*

Proof We show that $\mathcal{F} - \mathcal{F}^*$ is a small disturbance PUBST-ISS Lyapunov function. Consider $\mathcal{V} = \mathcal{F} - \mathcal{F}^*$ as a candidate PUBST-ISS-Lyapunov function. As we can see, $\mathcal{F} - \mathcal{F}^*$ is a size function with respect to x^*. Further, for interval $[t_0, t_f)$, By taking the time derivative of $\mathcal{V}$ along the dynamics of (4.20), we get

$$\begin{aligned} \dot{\mathcal{V}} &= (\nabla\mathcal{F})^\top \dot{x} \\ &= -\theta_1 (\nabla\mathcal{F})^\top P^\top \text{Sign}(\wp)\varphi(t, \wp) - \theta_1 \|\nabla\mathcal{F}\|^2 + \\ &\quad (\nabla\mathcal{F})^\top \Delta x(t), \end{aligned}$$

Using Cauchy-Schwarz inequality, we can further write,

$$\dot{\mathcal{V}} \leq -\theta_1 |\wp|^\top \varphi(t, \wp) - \theta_1 \|\nabla\mathcal{F}\|^2 + \|\nabla\mathcal{F}\| \|\Delta x\|,$$

where $|\wp| = (|\wp_1|, \ldots, |\wp_n|) \in \mathbb{R}^n$. Next, for any $i \in (1, .., n)$

$$\dot{\mathcal{V}} \leq -\theta_1 |\wp_i| \varphi_i - \theta_1 \|\nabla\mathcal{F}\|^2 + \|\nabla\mathcal{F}\| \|\Delta x\|,$$

Using Young's inequality, we get,

$$\begin{aligned} \dot{\mathcal{V}} &\leq -\theta_1 |\wp_i| \varphi_i - \theta_1 \|\nabla\mathcal{F}\|^2 + \frac{\|\nabla\mathcal{F}\|^2}{2} + \frac{\|\Delta x\|^2}{2} \\ &\leq -\theta_1 |\wp_i| \varphi_i + \frac{\|\Delta x\|^2}{2} \end{aligned}$$

Using Lemma 4.3, we have, $|\wp_i|^2 \geq \frac{\mu}{n}(\mathcal{F} - \mathcal{F}^*)$, and define, $\delta_d := \sqrt{\frac{\mu}{n}}$, where, $\mu > 0$, we can further write,

$$\dot{\mathcal{V}} \leq -\theta_1 \delta_d \sqrt{\mathcal{V}} \frac{(e^{\delta_d \sqrt{\mathcal{V}}} - 1)}{e^{\delta_d \sqrt{\mathcal{V}}}(t_f - t)} + \frac{\|\Delta x\|^2}{2},$$

Let $\hat{\mathcal{V}} = \delta_d \sqrt{\mathcal{V}}$. Then $\dot{\hat{\mathcal{V}}} = \frac{\delta_d^2}{2} \frac{\dot{\mathcal{V}}}{\hat{\mathcal{V}}}$. Therefore,

$$\dot{\hat{\mathcal{V}}} \leq -\hat{\theta} \frac{(e^{\hat{\mathcal{V}}} - 1)}{e^{\hat{\mathcal{V}}}(t_f - t)} + \frac{\|\Delta x\|^2}{2},$$

where $\hat{\theta} = \frac{\theta_1 \delta_d^2}{2} > 1$. Let $\frac{\|\Delta x\|^2}{2} \leq \gamma(\|\Delta x\|)$, where, $\gamma(\cdot)$ is a class $\mathcal{K}$ function. Then we can further write using Lemma 3.11

$$\dot{\hat{\mathcal{V}}} \leq -\beta(\hat{\mathcal{V}}, t_f - t) + \gamma(\|\Delta x\|) \tag{4.21}$$

where, $\beta(\cdot, \cdot)$ is a $\mathcal{PGKL}$ class function and $\gamma(\cdot)$ is a class $\mathcal{K}$ function. Next, we perform the analysis for $t \geq t_f$. By computing the time derivative of $\mathcal{V}$ along the dynamics of (4.9) for $t \geq t_f$, we get

$$\dot{\mathcal{V}} = -\theta_1 (\nabla \mathcal{F})^\top \wp(x) + (\nabla \mathcal{F})^\top \Delta x(t)$$

Using the property of P i.e. $P^\top P = P,\ P^\top = P$, Cauchy-Schwarz inequality, and Young's inequality, we get

$$\dot{\mathcal{V}} \leq -\theta_1 \|\wp\|^2 + \frac{\|\Delta x\|^2}{2}$$

using Lemma 4.3, we get

$$\dot{\mathcal{V}} \leq -\alpha_d(\mathcal{V}(x)) + \gamma(\|\Delta x\|) \tag{4.22}$$

where, $\alpha_d(\cdot)$ is a class $\mathcal{K}$ function. Combining (4.21) and (4.22), we get

$$\dot{\hat{\mathcal{V}}} \leq -\beta(\hat{\mathcal{V}}, t_f - t) - \alpha_d(\hat{\mathcal{V}}(x)) + \gamma(\|\Delta x\|),\ \forall\, t \geq 0.$$

If $\alpha_d(\hat{\mathcal{V}}(x)) \geq \gamma(\|\Delta x\|)$, then we get

$$\dot{\hat{\mathcal{V}}} \leq -\beta(\hat{\mathcal{V}}, t_f - t) - \alpha_d(\hat{\mathcal{V}}(x)),\ \ \forall\, t \geq 0. \tag{4.23}$$

Hence, according to Lemma 3.11, $\mathcal{V}$ is a small disturbance PUBST-ISS Lyapunov function. Further, according to Lemma 3.5, dynamics (4.20), is a small disturbance PUBST-ISS. □

4.6 Illustrative Examples

In the current section, we discuss problems based on constrained linear least squares and resource allocation using the proposed predefined-time convergent PGF techniques. Here the results are also compared with the FTS stable signed projected GF system proposed in [10].

4.6.1 Linear Least Squares

Consider the following problem:

$$\min_{x\in\mathbb{R}^4} \frac{1}{2}||Cx-d||^2$$
$$\text{subject to } Ax=b \tag{4.24}$$

where $C = \begin{bmatrix} 0.9501 & 0.7620 & 0.6153 & 0.4057 \\ 0.2311 & 0.4564 & 0.7919 & 0.9354 \\ 0.6068 & 0.0185 & 0.9218 & 0.9169 \\ 0.4859 & 0.8214 & 0.7382 & 0.4102 \\ 0.8912 & 0.4447 & 0.1762 & 0.8936 \end{bmatrix} \in \mathbb{R}^{4\times5}$, $d = \begin{bmatrix} 0.0578 \\ 0.3528 \\ 0.8131 \\ 0.0098 \\ 0.1388 \end{bmatrix} \in \mathbb{R}^{5\times1}$, $A = \begin{bmatrix} 3 & 5 & 7 & 9 \end{bmatrix} \in \mathbb{R}^{1\times4}$, $b = 4$. The optimal points of (4.24) is: $x^* = [0.017 \quad -0.594 \quad 0.513 \quad 0.369]^\top$. We first solve the Linear Least Squares (LLS) problem using the dynamics presented in [10] and compare its results from algorithm (4.9). For both the techniques, we take initial time as $t(0) = 0$ s and set $x(0) = [0 \quad 0 \quad 0 \quad 0.4444] \in \mathbb{X}_{feas}$. Figure 4.3 shows that state trajectories reach x^* around 0.12 s using the algorithm proposed in [10], whereas Fig. 4.4 shows the evolutions of states and an optimal solution is reached in predefined-time $t_f = 0.05$ s using dynamics (4.9).

4.7 Resource Allocation Problem (RAP)

Consider the economic dispatch problem with the following quadratic cost function: $f(\mathcal{P}_j) = a_j + b_j\mathcal{P}_j + c_j\mathcal{P}_j^2$, where a_j, b_j, c_j are cost coefficients.

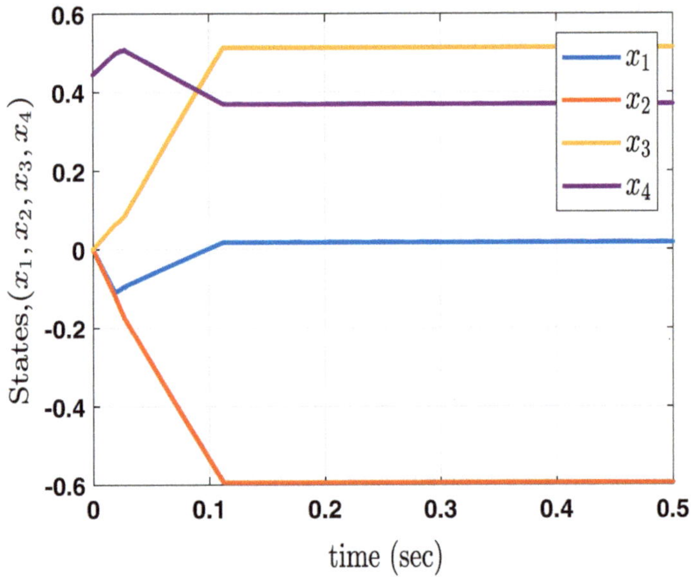

Fig. 4.3 States with finite-time convergence (FTC) using dynamics in [10] with **convergence time around 0.12 s**

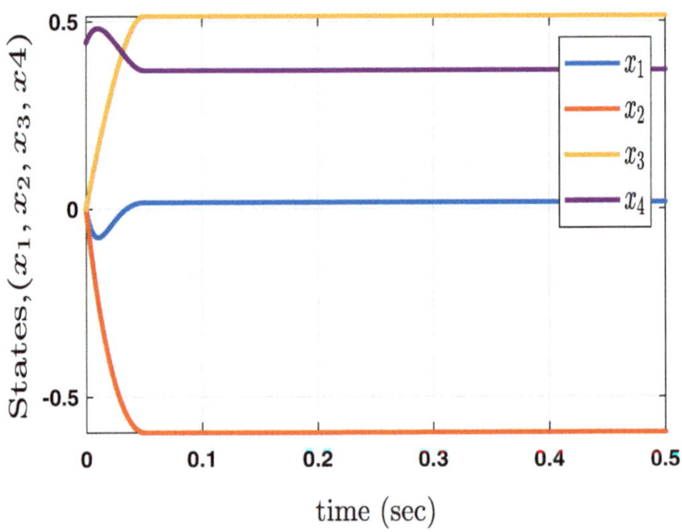

Fig. 4.4 States with PUBST convergence as per our proposed dynamics (4.9) with prior chosen **convergence time around 0.05 s**

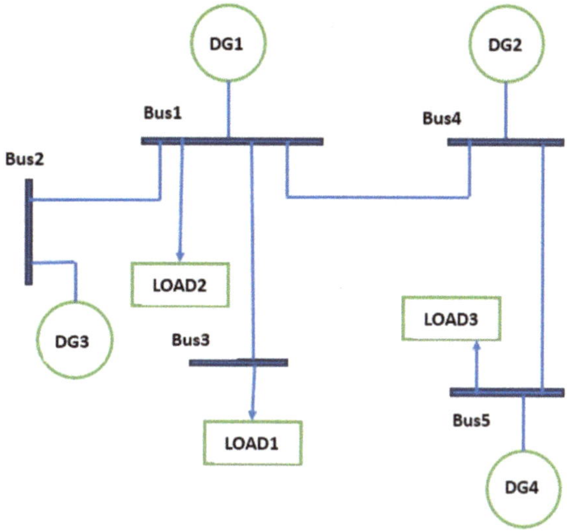

Fig. 4.5 Microgrid system with four generators

Table 4.1 Generators cost parameters

Generator	a_j	b_j	c_j
1	0.5	3	2
2	1.4	4	1
3	3	5	0.5
4	1	2	1.5

The layout of the Microgrid system is shown in Fig. 4.5. There are three loads and four generators in the system. The parameters of the generators are shown in Table 4.1. The total demand of the system is fixed as $d_0 = 145$. We take initial conditions as: $\mathcal{P}_1(0) = 40, \mathcal{P}_2(0) = 40, \mathcal{P}_3(0) = 40, \mathcal{P}_4(0) = 25$ which is inside the feasible set according to Proposition 4.1. The optimal power outputs of each generator are: $\mathcal{P}_1^* = 17.66, \mathcal{P}_2^* = 34.82, \mathcal{P}_3^* = 68.64, \mathcal{P}_4^* = 23.88$. Similar to the LLS example, here also, we first solve the RAP problem using the algorithm given in [10] and compare its results from algorithm (4.9). In Fig. 4.6, state trajectories reach their optimal value around 5 s, using the algorithm proposed in [10]. Here, we cannot choose the convergence time, whereas in Fig. 4.7, states converge to their optimal points within predefined-time $t_f = 0.2$ s, which is selected a priori, using proposed algorithm (4.9), with $\theta = 4$.

4.8 Conclusion

In this chapter, the predefined-time convergent gradient flow systems are proposed to solve continuous time optimization problems. Under the condition that the objective function is strongly convex or satisfies the P-L inequality. A predefined-

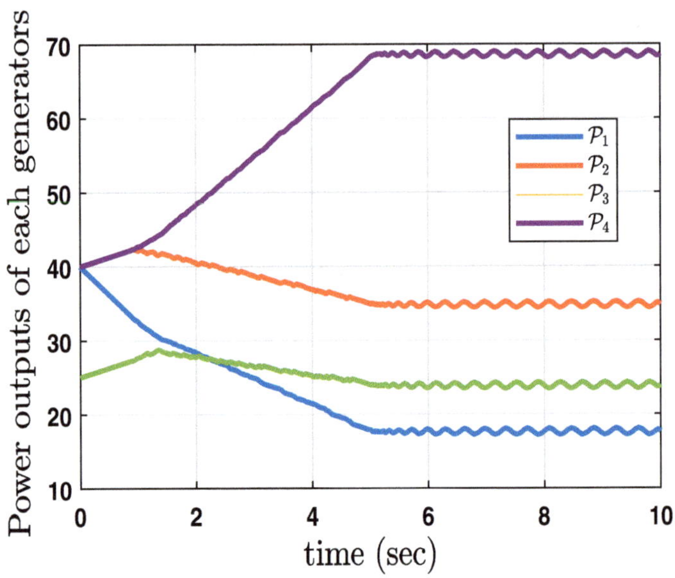

Fig. 4.6 Power outputs of generators with FTC as per the dynamics in [10] with **convergence time 5 s**

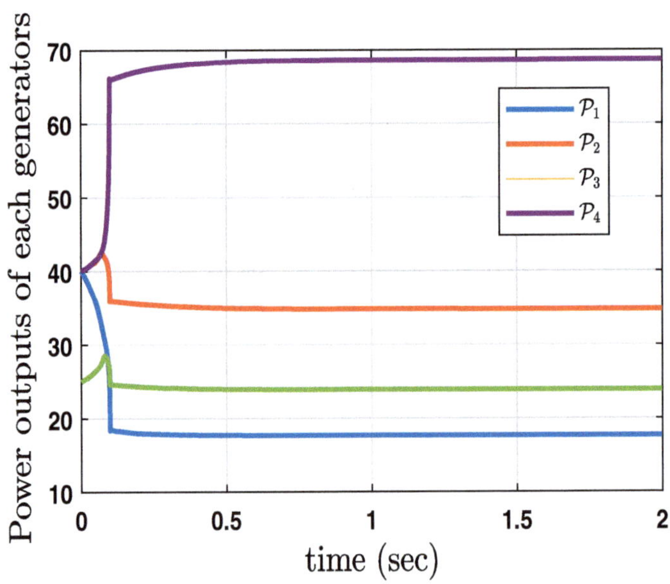

Fig. 4.7 Power outputs of generators with PUBST convergence as per our proposed dynamics (4.9) within prior chosen **convergence time 0.2 s**

time convergent projected gradient flow algorithm is proposed to solve convex optimization problems with LEC. Further, a robustness analysis using the idea of ISS is also discussed, considering perturbed PUBST based PGF dynamics. Next, the convergence analysis of the proposed dynamics are discussed using Lyapunov stability theory. The convergence time of proposed algorithms can be specified in advance and does not involve the initial value of the system and any parameters.

In the upcoming chapter, we propose modified GF dynamics to solve constrained TVO problems with a predefined-time convergence property.

References

1. Absil, P.-A., Kurdyka, K.: On the stable equilibrium points of gradient systems. Syst. Control Lett. **55**(7), 573–577 (2006)
2. Pal, A.K., Kamal, S., Yu, X., Nagar, S.K., Xiong, X.: Free-will arbitrary time consensus for multiagent systems. IEEE Trans. Cybern. **52**, 4636 (2020)
3. Jin, J., Chen, W., Chen, C., Chen, L., Tang, Z., Chen, L., Wu, L., Zhu, C.: A predefined fixed-time convergence ZNN and its applications to time-varying quadratic programming solving and dual-arm manipulator cooperative trajectory tracking. IEEE Trans. Ind. Inf. **19**(8), 8691–8702 (2022)
4. Wang, H., Tong, M., Zhao, X., Niu, B., Yang, M.: Predefined-time adaptive neural tracking control of switched nonlinear systems. IEEE Trans. Cybern. **53**, 6538 (2022)
5. Prasun, P., Pandey, S., Kamal, S., Ghosh, S., Singh, D., Ghosh, D.: Predefined upper bound of settling time based convergent gradient flow systems. In: IECON 2022–48th Annual Conference of the IEEE Industrial Electronics Society, pp. 1–6. IEEE (2022)
6. Pandey, S., Prasun, P., Kamal, S., Singh, D., Dinh, T.N.: Optimisation by predefined upper bound of settling time convergent projected gradient flows. Int. J. Syst. Sci., 1–14 (2025)
7. Pandey, S., Prasun, P., Kamal, S., Singh, D., Ghosh, D., Olaru, S.: Optimal trajectory tracking using Newton and Levenberg-Marquardt-like algorithms for constrained time-varying optimization. Optimal Control Appl. Methods **46**(4), 1800–1813 (2025)
8. Boyd, S.P., Vandenberghe, L.: Convex Optimization. Cambridge University Press (2004)
9. Strang, G.: Introduction to Linear Algebra. SIAM (2022)
10. Chen, F., Ren, W.: Sign projected gradient flow: A continuous-time approach to convex optimization with linear equality constraints. Automatica **120**, 109156 (2020)
11. Sontag, E.D.: Remarks on input to state stability of perturbed gradient flows, motivated by model-free feedback control learning. Syst. Control Lett. **161**, 105138 (2022)

Chapter 5
A Newton and Levenberg–Marquardt Framework for Optimal Trajectory Tracking in Constrained Time-Varying Optimization

Abstract This chapter develops continuous-time algorithms for tracking optimal trajectories in time-varying optimization problems within a predefined time. Newton-like and robust formulations are proposed to handle unknown time variations, while a Levenberg–Marquardt-like approach addresses singular or ill-conditioned Hessians. The framework is extended to constrained problems using barrier-based methods. Lyapunov-based analysis guarantees predefined-time convergence and robustness. Numerical results demonstrate improved tracking performance compared to existing methods.

5.1 Introduction

This chapter presents a CT dynamical framework for tracking the optimal trajectory of TVO problems with time-varying constraints within a predefined time, where the convergence time is specified a priori. A robust Newton-like scheme is developed to address scenarios in which the exact rate of variation of the objective function gradient is unavailable. Furthermore, a Levenberg–Marquardt-type approach is proposed to handle cases where the Hessian matrix is singular or ill-conditioned. The convergence properties of the proposed methods are rigorously established using Lyapunov-based analysis. Numerical simulations on representative TVO problems demonstrate the effectiveness of the proposed framework.

The rest of the chapter is structured as follows: In Sect. 5.2, problem statement is discussed. Section 5.3 presents the main result for unconstrained time-varying optimization problems. In Sect. 5.4, the results discussed in Sect. 5.3 are extended for constrained TVO. Numerical examples are illustrated in Sect. 5.5 based on the results discussed in Sects. 5.3 and 5.4.

S. Kamal, S. Pandey, *A Continuous-Time Perspective on Accelerated Optimization and Control*, Studies in Systems, Decision and Control 693,
https://doi.org/10.1007/978-3-032-31645-5_5

5.2 Problem Statement

Consider the following TVO problem

$$\begin{aligned} x^*(t) := & \operatorname*{argmin}_{x\in\mathbb{R}^n} \mathcal{F}(t,x) \\ \text{subject to} \quad & g_i(t,x) \le 0, \quad i \in \{1,2,\dots,r\} \end{aligned} \tag{5.1}$$

where $\mathcal{F}: \mathbb{R}_{\ge 0}\times\mathbb{R}^n \to \mathbb{R}$, $g = [g_1, g_2, \dots, g_r]^\top$ with all g_i being convex functions in $x \in \mathbb{R}^n$, $\forall t \ge 0$ and fulfills following Assumptions.

Assumption 5

1. *The objective function $\mathcal{F}: \mathbb{R}_{\ge 0}\times\mathbb{R}^n \to \mathbb{R}$ is twice continuously differentiable with respect to x, continuously differentiable with respect to t, and convex with respect to x.*
2. *A unique optimal solution $x^*(t)$ exists.*

Assumption 6 (Polyak-Lojasiewicz Inequality (See [1]))

$$w(\mathcal{F}(t,x) - \mathcal{F}^*(t))^p \le \|\nabla\mathcal{F}(t,x)\| \tag{5.2}$$

where $\mathcal{F}^* := \min_{z\in\mathbb{R}^n} \mathcal{F}(t,x)$, $0 < p < 1$ *and* $w > 0$.

Assumption 7 *The inequality constraint functions $g_i(t,x)$ are twice continuously differentiable with respect to x and continuously differentiable with respect to t.*

Assumption 8 (Constraint Qualification) *There exists atleast one $x(t) \in \mathbb{R}^n$ such that $g_i(t,x) < 0$, $\forall i \in \{1,2,\dots r\}$ and $\forall t \ge 0$.*

5.3 Time-Varying Unconstrained Convex Optimization

Consider the following unconstrained TVO problem

$$x^*(t) := \operatorname*{argmin}_{x\in\mathbb{R}^n} \mathcal{G}(t,x), \tag{5.3}$$

Remark 5.1 For TVO problem (5.3), Assumption 5 is an important condition for the convergence of gradient-based methods to an optimal solution [2]. In Karimi et al. [1], it is noted that Assumption 6 is the weakest condition among similar conditions commonly used in the literature to demonstrate convergence in gradient-based algorithms.

Lemma 5.1 (See [2]) *The continuously differentiable convex function $\mathcal{G}: \mathbb{R}_{\ge 0}\times\mathbb{R}^p \to \mathbb{R}$ is minimized at solution $x^*(t)$ if and only if $\nabla\mathcal{G}(t,x^*) = 0$, $\forall t$.*

The optimal trajectory $x^*(t)$ satisfies the first order optimality condition $\nabla\mathcal{G}(t,x^*) = 0, \quad \forall t \geq 0$, then time derivative of this condition will also be zero which results in $\nabla^2\mathcal{G}(t,x)\dot{x}^*(t) + \frac{\partial\nabla\mathcal{G}(t,x)}{\partial t} = 0$. Solving for $\dot{x}^*(t)$ leads to following dynamical system:

$$\dot{x}^*(t) = -\nabla^2\mathcal{G}(t,x^*))^{-1}\frac{\partial\nabla\mathcal{G}(t,x^*)}{\partial t}, \tag{5.4}$$

If the optimal solution $x^*(t)$ was known for some $t_0 > 0$, the system in (5.4) could be used to track the evolution of $x^*(t)$. As we do not have access to $x^*(t)$ at any point in time, we propose a predefined-time convergent dynamics to solve the TVO problem in a priori chosen time, utilizing the idea of prediction and correction.

In the next section, we discuss predefined-time convergent dynamics to solve TVO problems. The main result of this chapter discusses TVO problems for which $\underset{t\to t_f}{lim}\|x(t) - x^*(t)\| = 0$, where $t_f < \infty$, that is chosen a priori. The convergence of the trajectories based on the predefined-time dynamics to the optimal solution $x^*(t)$ also signifies $\underset{t\to t_f}{lim}\|\nabla\mathcal{G}(t,x)\| = 0$. First, we consider TVO problems (5.3), where $\mathcal{G}(t,x)$ is given so $\nabla\mathcal{G}(t,x)$ is known $\forall t$, and aim is to converge to optimal solution $x^*(t)$ within priori chosen time t_f.

5.3.1 *Predefined-Time Convergent Dynamics for Time-Varying Optimization (PTC-TVO)*

In this subsection, we present two predefined-time convergent dynamical systems to solve unconstrained TVO problems as (5.3). One is based on the assumption that $\nabla^2\mathcal{G}$ is invertible and second if we do not consider $\nabla^2\mathcal{G}$ to be invertible. We revise Eq. (5.4) by adding a correction term to propose dynamics that track the optimal trajectory in a predefined time.

5.3.1.1 Newton-like PTC-TVO

Let $\nabla^2\mathcal{G}$ is invertible then in order to track $x^*(t)$ of (5.3) within predefined-time, we propose the following dynamics

$$\dot{x}(t) = \begin{cases} -(\nabla^2\mathcal{G}(t,x))^{-1}\left(\vartheta\psi(t,x) + \frac{\partial\nabla\mathcal{G}(t,x)}{\partial t}\right), \text{for } t_0 \leq t < t_f \\ -(\nabla^2\mathcal{G}(t,x))^{-1}\left(\vartheta\nabla\mathcal{G}(t,x) + \frac{\partial\nabla\mathcal{G}(t,x)}{\partial t}\right), \text{ for } t \geq t_f, \end{cases} \tag{5.5}$$

where $x(t) \in \mathbb{R}^n$, $\vartheta \in \mathbb{R}$ and satisfies $\vartheta > 1$, $\psi(t,x) = (\psi_1, \psi_2, \ldots, \psi_n) \in \mathbb{R}^{n\times 1}$, $\psi_i := \frac{(e^{\nabla\mathcal{G}_{x_i}(t,x)} - 1)}{e^{\nabla\mathcal{G}_{x_i}(t,x)}(t_f - t)}$, for $1 \leq i \leq n$.

Theorem 5.1 ([3]) *Consider the dynamics* (5.5). *Suppose that* $\mathcal{G} : \mathbb{R}_{\geq 0} \times \mathbb{R}^n \to \mathbb{R}$ *in* (5.3) *has invertible Hessian, satisfies Assumption 5. There exist a continuously differentiable function* $V : [t_0, \infty) \times \mathbb{R}^n \to \mathbb{R}$, *satisfying* $\gamma_1(x) \leq V(t, x) \leq \gamma_2(x)$, $\forall x \in \mathbb{R}^n \setminus \{0\}$, $\forall t \in [t_0, \infty)$, $\gamma_1(x)$ *and* $\gamma_2(x)$ *be continuous positive definite functions,* $V(t, x^*) = 0$ *and for* $V \neq 0$: $\dot{V} \leq 0, \forall t \geq t_f$ *and* $\dot{V} \leq \frac{-\vartheta(e^V-1)}{e^V(t_f-t)}$, $\forall t \in [t_0, t_f)$. *Then* $x(t) \to x^*(t)$ *as* $t \to t_f$. *This signifies that* $x(t)$ *converges to the optimal solution* $x^*(t)$ *in a priori chosen time.*

Proof Let us select the Lyapunov function candidate

$$V = \|\nabla \mathcal{G}(t, x)\|^2, \tag{5.6}$$

The derivative of the V along the dynamics (5.5) for $t \in [t_0, t_f)$

$$\begin{aligned}
\dot{V}(x) =& 2(\nabla \mathcal{G}(t, x))^\top \left(\nabla^2 \mathcal{G}(t, x)\dot{x} + \frac{\partial \nabla \mathcal{G}(t, x)}{\partial t} \right) \\
=& 2\nabla \mathcal{G}^\top \left(\nabla^2 \mathcal{G}\Big(- (\nabla^2 \mathcal{G})^{-1} \Big(\vartheta \psi + \frac{\partial \nabla \mathcal{G}}{\partial t} \Big)\Big) + \frac{\partial \nabla \mathcal{G}}{\partial t} \right) \\
=& - 2\vartheta (\nabla \mathcal{G}(t, x)^\top \psi(t, x) \qquad (5.7)\\
\leq& - 2\vartheta (\nabla \mathcal{G}_{x_i}(t, x))\psi_i(t, x) \qquad \text{for any i} \\
\leq& - 2\vartheta \frac{|\nabla \mathcal{G}_{x_i}(t, x)|(e^{|\nabla \mathcal{G}_{x_i}(t,x)|} - 1)}{e^{|\nabla \mathcal{G}_{x_i}(t,x)|}(t_f - t)} \qquad (5.8)
\end{aligned}$$

Since, $V \leq n(\max\{|\nabla \mathcal{G}_1|, |\nabla \mathcal{G}_2|, \ldots, |\nabla \mathcal{G}_n|\}^2)$ and thus, $\sqrt{V/n} \leq \max\{|\nabla \mathcal{G}_1|, |\nabla \mathcal{G}_2|, \ldots, |\nabla \mathcal{G}_n|\}$. We assume that at a particular time, $\max\{|\nabla \mathcal{G}_1|, |\nabla \mathcal{G}_2|, \ldots, |\nabla \mathcal{G}_n|\}$ returns $|\nabla \mathcal{G}_i|$, then, $\sqrt{V/n} \leq |\nabla \mathcal{G}_i|$. Using this fact in (5.8)

$$\dot{V}(x) \leq -2\vartheta \frac{\sqrt{V/n}(e^{\sqrt{V/n}} - 1)}{e^{\sqrt{V/n}}(t_f - t)}. \tag{5.9}$$

Let $\hat{V} = \sqrt{V/n}$, hence, $\dot{\hat{V}} = \frac{\dot{V}}{2n\hat{V}}$, so (5.9) is simplified and yields,

$$\dot{\hat{V}} \leq -\frac{\hat{\vartheta}(e^{\hat{V}} - 1)}{e^{\hat{V}}(t_f - t)} \tag{5.10}$$

where $\hat{\vartheta} = \frac{\vartheta}{n} > 1$. Thus, inequality (5.10) represents PUBST based stability according to Lemma 3.2, which means that $\|\nabla \mathcal{G}(t, x)\| \to 0$ as $t \to t_f$, leading to the convergence of trajectory of (5.5) to the optimal solution $x^*(t)$ of $\mathcal{G}(t, x)$ in a priori chosen time. Next, the time derivative of the V along (5.5) for $t \geq t_f$ gives

$$\begin{aligned}\dot{V}(x) =&2(\nabla \mathcal{G}(t,x))^\top \left(\nabla^2 \mathcal{G}(t,x)\dot{x} + \frac{\partial \nabla \mathcal{G}(t,x)}{\partial t}\right)\\ =&2\nabla \mathcal{G}^\top \left(\nabla^2 \mathcal{G}\Big(-(\nabla^2 \mathcal{G})^{-1}\Big(\vartheta \nabla \mathcal{G} + \frac{\partial \nabla \mathcal{G}}{\partial t}\Big)\Big) + \frac{\partial \nabla \mathcal{G}}{\partial t}\right)\\ =&-2\vartheta (\nabla \mathcal{G}(t,x))^\top \nabla \mathcal{G}(t,x)\\ =&-2\vartheta ||\nabla \mathcal{G}(t,x)||^2\\ =&-2\vartheta V\end{aligned}$$

So, for $t \geq t_f$, the equilibrium point of system (5.5) is asymptotically stable, therefore, trajectories of (5.5) will maintain $x^*(t)$. Hence, $x(t) \to x^*(t)$ as $t \to t_f$. □

Remark 5.2 In (5.5), it is required to know the true information of $\frac{\partial \nabla \mathcal{G}(t,x)}{\partial t}$.

5.3.1.2 Robust Newton-like PTC-TVO

In the previous subsection, we require the exact knowledge of $\frac{\partial \nabla \mathcal{G}(t,x)}{\partial t}$. Here, we assume that $\frac{\partial \nabla \mathcal{G}(t,x)}{\partial t}$ is unknown, but $\exists \ell \in \mathbb{R}_{>0}$ such that $\left\|\frac{\partial \nabla \mathcal{G}(t,x)}{\partial t}\right\| \leq \ell$, then in order to track $x^*(t)$ of (5.3) within predefined-time, we propose the following dynamics

$$\dot{x}(t) = \begin{cases} -(\nabla^2 \mathcal{G}(t,x))^{-1}\left(\vartheta \psi(t,x) + \ell \frac{\nabla \mathcal{G}(t,x)}{\|\nabla \mathcal{F}(t,x)\|}\right), & t_0 \leq t < t_f \\ -(\nabla^2 \mathcal{G}(t,x))^{-1}\left(\vartheta \nabla \mathcal{F}(t,x) + \ell \frac{\nabla \mathcal{G}(t,x)}{\|\nabla \mathcal{F}(t,x)\|}\right), & t \geq t_f, \end{cases} \tag{5.11}$$

where $x(t) \in \mathbb{R}^n$, $\vartheta \in \mathbb{R}$ and satisfies $\vartheta > 1$, $\psi(t,x) = (\psi_1, \psi_2, \ldots, \psi_n) \in \mathbb{R}^{n\times 1}$, $\psi_i := \frac{(e^{\nabla \mathcal{G}_{x_i}(t,x)} - 1)}{e^{\nabla \mathcal{G}_{x_i}(t,x)}(t_f - t)}$, for $1 \leq i \leq n$.

Theorem 5.2 ([3]) *Consider the dynamics* (5.11). *Suppose that* $\mathcal{G} : \mathbb{R}_{\geq 0} \times \mathbb{R}^n \to \mathbb{R}$ *in* (5.3) *has invertible Hessian with* $\left\|\frac{\partial \nabla \mathcal{G}}{\partial t}\right\| \leq \ell$, $\ell \in \mathbb{R}_+$ *satisfies Assumption 5 and* $\exists$ *a continuously differentiable function* $V : [t_0, \infty) \times \mathbb{R}^n \to \mathbb{R}$, *satisfying* $\gamma_1(x) \leq V(t,x) \leq \gamma_2(x)$, $\forall x \in \mathbb{R}^n \setminus \{0\}, \forall t \in [t_0, \infty)$, $\gamma_1(x)$ *and* $\gamma_2(x)$ *be continuous positive definite functions,* $V(t,x^*) = 0$ *and for* $V \neq 0$*:* $\dot{V} \leq 0, \forall t \geq t_f$ *and* $\dot{V} \leq \frac{-\vartheta(e^V - 1)}{e^V(t_f - t)}, \forall t \in [t_0, t_f)$ *for* (5.11). *This signifies that* $x(t)$ *converges to the optimal solution* $x^*(t)$ *in a priori chosen time.*

Proof Let us select the Lyapunov function candidate

$$V = \|\nabla \mathcal{G}(t,x)\|^2, \tag{5.12}$$

The derivative of the V along the dynamics (5.11) for $t \in [t_0, t_f)$

$$\begin{aligned}\dot{V}(x) =& 2(\nabla \mathcal{G}(t,x))^\top \left(\nabla^2 \mathcal{G}(t,x)\dot{x} + \frac{\partial \nabla \mathcal{G}(t,x)}{\partial t}\right)\\ =& 2(\nabla \mathcal{G})^\top \left(-\vartheta \psi - \ell \frac{\nabla \mathcal{G}}{\|\nabla \mathcal{G}\|} + \frac{\partial \nabla \mathcal{G}}{\partial t}\right)\end{aligned}$$

Using Cauchy—Schwarz inequality, we can further write,

$$\leq -2\vartheta \nabla \mathcal{G}^\top \psi - 2\ell \|\nabla \mathcal{G}\| + 2\|\nabla \mathcal{G}\| \left\|\frac{\partial \nabla \mathcal{G}}{\partial t}\right\|$$

Since, $\left\|\frac{\partial \nabla \mathcal{G}}{\partial t}\right\| \leq \ell$, then further we get,

$$\dot{\mathcal{V}}(z) \leq -2\vartheta \nabla \mathcal{G}^\top \psi$$

The above inequality is similar to inequality (5.7). Then proceeding in a similar way as in the proof of Theorem 5.1, we get inequality (5.10) which represents PUBST-based stability, implying the trajectory of (5.11) converges to the optimal solution of $\mathcal{G}(t,x)$ in a priori chosen time. Next for $t > t_f$, proceeding in a similar way as for $t \in [t_0, t_f)$, we get $\dot{\mathcal{V}} \leq 2\vartheta \mathcal{V}$, hence trajectories of (5.11) will maintain $x^*(t)$. Hence, $x(t) \to x^*(t)$ as $t \to t_f$. □

Computational Cost We can discuss the convergence rates of Euler discretization to calculate complexity considering the assumption of Lipschitz smoothness. Let us consider the Euler discretized version of the proposed Robust Newton-like method for PUBSTC-TVO dynamics (5.11). To simplify the computation of complexity we can write $\psi(t,x) := \frac{\nabla \mathcal{G}(t,x)}{t_f - t}$. Let, $\mathfrak{K} = t_f/h$. Hence, the considered discretized version of the proposed dynamics, taking h as practical step size, can be written as:

$$x(k+1) = \begin{cases} x(k) - h\left(\frac{\vartheta}{\mathfrak{K}-k} + \frac{\ell}{\|\mathcal{G}(k,x)\|}\right)(\nabla^2 \mathcal{G}(k,x))^{-1}\nabla \mathcal{G}(k,x) & \text{for } 0 \leq k < \mathfrak{K} \\ x(k) - h\left(\vartheta + \frac{\ell}{\|\mathcal{G}(k,x)\|}\right)(\nabla^2 \mathcal{G}(k,x))^{-1}\nabla \mathcal{G}(k,x) & \text{for } \mathfrak{K} \leq k < \infty \end{cases} \tag{5.13}$$

Let, $\gamma_{1k} := \left(\frac{\vartheta}{\mathfrak{K}-k} + \frac{\ell}{\|\mathcal{G}(k,x)\|}\right)$, and $\gamma_{2k} := \left(\vartheta + \frac{\ell}{\|\mathcal{G}(k,x)\|}\right)$ and $\gamma_{1k}, \gamma_{2k} \in \mathbb{R}_{>0}$.

We need to introduce an extra assumption of L-Lipschitz smooth, as given below:

Assumption (Lipschitz Smoothness of Order q) We assume the function $\mathcal{G}$ is L-Lipschitz smooth of order $q \in (1, 2]$, i.e., for any $x, y \in \mathbb{R}^n$,

$$\|\nabla \mathcal{G}(k, y) - \nabla \mathcal{G}(k, x)\| \leq L\|y - x\|^{q-1} \tag{5.14}$$

The above assumption is also called (L, q) Holder continuity, it will lead to the following property:

$$\mathcal{G}(k, y) \leq \mathcal{G}(k, x) + \langle \nabla \mathcal{G}(k, x), y - x \rangle + \frac{L}{q}\|y - x\|^q, \tag{5.15}$$

When $q = 2$, the function will be Lipschitz smooth.

Considering the following assumptions, for our case, we can calculate iteration complexity as follows:

$$\begin{aligned} \mathcal{G}(k+1, x(k+1)) \leq \mathcal{G}(k, x(k)) &+ \langle \nabla \mathcal{G}(k, x(k)), x(k+1) - x(k) \rangle \\ &+ \frac{L}{q}\|x(k+1) - x(k)\|^q, \end{aligned} \tag{5.16}$$

Using algorithm (5.13), for $0 \leq k < \mathfrak{K}$ and considering the Lipschitz continuity of the gradient and for $q = 2$

$$\begin{aligned} \mathcal{G}(k+1, x(k+1)) &\leq \mathcal{G}(k, x(k)) - \frac{\gamma_{1k} h}{L}\|\nabla \mathcal{G}\|^2 + \frac{\gamma_{1k}^2 h^2}{2L}\|\nabla \mathcal{G}\|^2 \\ &= \mathcal{G}(k, x(k)) - \frac{\gamma_{1k} h}{L}(1 - \gamma_{1k} h/2)\|\nabla \mathcal{G}\|^2 \end{aligned}$$

Further, using the Assumption 2 (PL inequality): $\|\nabla \mathcal{G}\|^2 \geq \alpha(\mathcal{G} - \mathcal{G}^*)$, where, $\alpha > 0$, we can further write

$$\mathcal{G}(k+1, x(k+1)) \leq \mathcal{G} - \frac{\gamma_{1k} h \alpha}{L}(1 - \gamma_{1k} h/2)(\mathcal{G} - \mathcal{G}^*(k))$$

Subtracting $\mathcal{G}^*$ on both sides and $\kappa := (1 - 0.5h\gamma_{1k})$, then

$$\mathcal{G}(k+1, x(k+1)) - \mathcal{G}^*(k) \leq \left(1 - \frac{\gamma_{1k} h \alpha \kappa}{L}\right)(\mathcal{G}(k, x(k)) - \mathcal{G}^*(k))$$

Then for $\mathfrak{K}$ iterations with a fixed step size h

$$\mathcal{G}(\mathfrak{K}, x(\mathfrak{K})) - \mathcal{G}^*(k) \leq \left(1 - \frac{\gamma_{1k} h \alpha \kappa}{L}\right)^{\mathfrak{K}}(\mathcal{G}(0, x(0)) - \mathcal{G}^*(k))$$

By the inequality, $e^{-x} \geq 1 - x$ we can get that the corresponding iteration complexity is $O\left(\left(\frac{\gamma_{1k} h\alpha\kappa}{L}\right)^{-1} ln\frac{1}{\epsilon}\right)$ for ϵ-closeness of $x(k)$ to $x^*(k)$. Similarly, if we run the algorithm for K iterations then for $\mathfrak{K} \leq k < K$, the corresponding iteration complexity is $O\left(\left(\frac{\gamma_{2k} h\alpha\kappa}{L}\right)^{-1} ln\frac{1}{\epsilon}\right)$.

5.3.1.3 Levenberg–Marquardt–like PTC-TVO without Invertible Hessian

Let inverse of $\nabla^2 \mathcal{G}$ is singular or near singular then the Newton-like technique is not well defined. Suppose $(\nabla^2 \mathcal{G} + \wp I_n)$ is invertible for some damping factor $\wp > 0$ then in order to track $x^*(t)$ of (5.3) in predefined-time, we propose the following dynamics:

$$\dot{x}(t) = \begin{cases} -(\nabla^2 \mathcal{G}(t,x) + \wp I_n)^{-1} \left(\vartheta \psi(t,x) + \frac{\partial \nabla \mathcal{G}(t,x)}{\partial t}\right), & \text{for } t_0 \leq t < t_f, \\ -(\nabla^2 \mathcal{G}(t,x) + \wp I_n)^{-1} \left(\vartheta \nabla \mathcal{G}(t,x) + \frac{\partial \nabla \mathcal{G}(t,x)}{\partial t}\right), & \text{for } t \geq t_f, \end{cases} \tag{5.17}$$

where $x(t) \in \mathbb{R}^n$, $\vartheta, \wp \in \mathbb{R}$ and satisfies $\vartheta > 1$, $\psi(t,x) = (\psi_1, \psi_2, \ldots, \psi_n) \in \mathbb{R}^{n \times 1}$, $\psi_i := \frac{(e^{\nabla \mathcal{G}_{x_i}(t,x)} - 1)}{e^{\nabla \mathcal{G}_{x_i}(t,x)} (t_f - t)}$, for $1 \leq i \leq n$ and consider the dynamics (5.17) as a perturbed version of (5.5) with input $\wp$.

Remark 5.3 One of the methods for calculating the damping factor at a particular instant includes taking the negative of the minimum eigenvalue of the Hessian matrix at that instant.

Theorem 5.3 ([3]) *Consider the dynamics* (5.17). *Suppose that* $\mathcal{G} : \mathbb{R}_{\geq 0} \times \mathbb{R}^n \to \mathbb{R}$ *in* (5.3) *satisfies Assumptions 5 and 6 with* $p = 1/2$ *and* $\exists$ *a PUBST-ISS Lyapunov function* $V : [t_0, \infty) \times \mathbb{R}^n \to \mathbb{R}$ *for* (5.17) *then dynamics* (5.17) *is PUBST-ISS.*

Proof Let us select the Lyapunov function candidate

$$V = \|\nabla \mathcal{G}(t,x)\|^2 + 2\wp(\mathcal{G}(t,x) - \mathcal{G}^*(t)) \tag{5.18}$$

The derivative of the V along the dynamics (5.17) gives

$$\begin{aligned} \dot{V} =& 2\nabla \mathcal{G}^\top \left(\nabla^2 \mathcal{G}\dot{x} + \frac{\partial \nabla \mathcal{G}}{\partial t}\right) + 2\wp \nabla \mathcal{G}^\top \dot{x} - 2\wp \dot{\mathcal{G}}^* \qquad (5.19) \\ =& 2\nabla \mathcal{G}^\top \left((\nabla^2 \mathcal{G} + \wp I_n)\Big(- (\nabla^2 \mathcal{G} + \wp I_n)^{-1}\Big(\vartheta \psi(t,x) + \frac{\partial \nabla \mathcal{G}}{\partial t}\Big)\Big) + \frac{\partial \nabla \mathcal{G}}{\partial t}\right) - 2\wp \dot{\mathcal{G}}^* \\ =& -2\vartheta (\nabla \mathcal{G}(t,x))^\top \psi(t,x) - 2\wp \dot{\mathcal{G}}^* \end{aligned}$$

Taking the worst case, we assume $\sup_{t\geq 0}(\dot{\mathcal{G}}^*(t)) < -a,\ a \in \mathbb{R}_{>0}$, then further we can write,

$$\dot{V} \leq -2\vartheta \frac{|\nabla \mathcal{G}_{x_i}(t,x)|(e^{|\nabla \mathcal{G}_{x_i}(t,x)|} - 1)}{e^{|\nabla \mathcal{G}_{x_i}(t,x)|}(t_f - t)} + 2a\wp, \qquad \text{for any i} \tag{5.20}$$

Next, using PL inequality with $p = 1/2$, we get $V \leq \left(1 + \frac{2\wp}{w}\right) \|\nabla \mathcal{G}\|^2$. Since, $V \leq n(1 + (2\wp/w))(\max\{|\nabla \mathcal{G}_1|, |\nabla \mathcal{G}_2|, \ldots, |\nabla \mathcal{G}_n|\}^2)$. We assume that at a particular time, $\max\{|\nabla \mathcal{G}_1|, |\nabla \mathcal{G}_2|, \ldots, \|\nabla \mathcal{G}_n|\}$ returns $|\nabla \mathcal{G}_i|$ and $n(1 + (2\wp/w)) = c$ then, $\sqrt{V/c} \leq |\nabla \mathcal{G}_i|$. Using this fact in (5.20) gives, $\dot{V} \leq -2\vartheta \frac{\sqrt{V/c}(e^{\sqrt{V/c}}-1)}{e^{\sqrt{V/c}}(t_f-t)} + 2a\wp$. Next, let $\hat{V} = \sqrt{V/c}$, hence, $\dot{\hat{V}} = \frac{\dot{V}}{2c\hat{V}}$. Further

$$\dot{\hat{V}} \leq -\frac{\hat{\vartheta}(e^{\hat{V}} - 1)}{e^{\hat{V}}(t_f - t)} + \gamma(|\wp|) \tag{5.21}$$

where γ is a class $\mathcal{K}$ function and suppose $\hat{\vartheta} = \frac{\vartheta}{c} > 1$. Next, for $t > t_f$, after Eq. (5.19) along the dynamics (5.17) proceeding in the similar way as for $t \in [t_0, t_f)$, we get $\dot{V} \leq -2\vartheta V + \gamma(|\wp|)$ and combining it with (5.21), we get $\dot{V} \leq -\beta(\|x_0\|, t_f - t) - \alpha(\|x\|) + \gamma(\|\wp\|)$, where β is $\mathcal{KL}$ class function and α, γ are the class $\mathcal{K}$ functions. Hence $V(t,x)$ satisfies the PUBST-ISS Lyapunov function condition thus from Lemma 3.5, system (5.17) is PUBST-ISS. □

Remark 5.4 The damping term $\wp$ functions as a regularizer, ensuring that $(\nabla^2 \mathcal{G}(t,x) + \wp I_n)$ remains positive definite even when the Hessian is singular, thereby guaranteeing its invertibility. The adaptive nature of $\wp$ influences the algorithm's behavior: $\wp$ is large (when the Hessian is singular), then the algorithm resembles a Gradient-like

$$\text{PTC-TVO algorithm: } \dot{x}(t) = \begin{cases} -(\wp I_n)^{-1}\left(\vartheta\psi(t,x) + \frac{\partial \nabla \mathcal{G}(t,x)}{\partial t}\right), & \text{for } t_0 \leq t < t_f, \\ -(\wp I_n)^{-1}\left(\vartheta\nabla \mathcal{G}(t,x) + \frac{\partial \nabla \mathcal{G}(t,x)}{\partial t}\right), & \text{for } t \geq t_f, \end{cases}$$

whereas $\wp$ is small (when the Hessian is well-conditioned), the algorithm exhibits behavior similar to the proposed Newton-like PTC-TVO algorithm (5.5).

5.4 Time-Varying Constrained Convex Optimization

In this section, we extend the results of Sect. 5.3.1 for constrained TVO problems with inequality constraints.

5.4.1 PTC-TVO with Inequality Constraints

Consider the constrained TVO problem

$$x^*(t) := \underset{x \in \mathbb{R}^n}{\text{argmin}}\, \mathcal{F}(t, x)$$
$$\text{subject to } g_i(t, x) \leq 0, \quad i \in \{1, 2, \ldots, r\} \tag{5.22}$$

where $g = [g_1, g_2, \ldots, g_r]^\top$ with all g_i being convex functions in $x \in \mathbb{R}^n, \ \forall t \geq 0$ and fulfills Assumptions 5, 6, 7, and 8.

To include the constraints in the objective function, we introduce a vector of Lagrange multipliers $\lambda \in \mathbb{R}^r_{\geq 0}$ and $\mathcal{L} : \mathbb{R}_{\geq 0} \times \mathbb{R}^n \times \mathbb{R}^r_{\geq 0} \to \mathbb{R}$ associated with TVO in (5.22) is

$$\mathcal{L}(t, x, \lambda) := \mathcal{F}(t, x) + \sum_{i=1}^{r} \lambda_i g_i(t, x) \tag{5.23}$$

The dual function is $\mathfrak{T}^*(t) := \underset{\lambda \in \mathbb{R}^r}{\arg\max}\, \mathfrak{G}(t, \lambda)$, where $\mathfrak{G}(t, \lambda) := \underset{x \in \mathbb{R}^n}{\min}\, \mathcal{L}(t, z, \lambda)$.

The necessary and sufficient KKT conditions [2] for optimality under Assumptions 5, 7, and 8, $\forall t \geq 0$ are

$$\nabla \mathcal{F}(t, x^*) + \sum_{i=1}^{r} \lambda_i^* \nabla g_i(t, x^*) = 0 \tag{5.24}$$

$$\lambda_i^* g_i(t, x^*) = 0, \tag{5.25}$$

$$\lambda_i^* \geq 0 \tag{5.26}$$

$$g_i(t, x^*) \leq 0, \ \ \forall i \in \{1, 2, \ldots, r\} \tag{5.27}$$

Next, we formulate perturbed KKT conditions, where we modify (5.25) as $\lambda_i^* g_i(t, \hat{x}^*) = -\frac{1}{\sigma} \implies \lambda_i^* = -\frac{1}{\sigma g_i(t, \hat{x}^*)}$, where $\sigma : \mathbb{R}_{\geq 0} \to \mathbb{R}_{>0}$ is defined as a barrier parameter and $(1/\sigma) \to 0$ as $\hat{x}(t) \to x^*(t)$. Next, upon substituting the obtained λ_i^* from perturbed KKT condition in (5.24), we get

$$\nabla(\mathcal{F}(t, \hat{x}^*) - \frac{1}{\sigma} \sum_{i=1}^{r} log(-g_i(t, x^*)) = 0 \tag{5.28}$$

which also satisfies KKT conditions (5.26) and (5.27). Next, we define the approximate optimal trajectory and Lagrangian $\mathcal{L} : \mathbb{R}_{\geq 0} \times \mathbb{R}^n \times \mathbb{R}_{>0} \times \mathbb{R}_{\geq 0} \to \mathbb{R}$ associated with (5.22) using $\sigma(t)$ and $\delta(t)$, where $\delta(t) : \mathbb{R}_{\geq 0} \to \mathbb{R}_{\geq 0}$ is defined as a slack function. It is to be noted that the $\delta(t)$ enlarges the initial feasible set.

$$\hat{x}^*(t) := \arg\min_{x\in\tilde{D}} \mathcal{F}(t,x) - \frac{1}{\sigma(t)}\sum_{i=1}^{r}\log(\delta(t) - g_i(t,x))$$

$$\tilde{\mathcal{L}}(t,x,\sigma,\delta) := \mathcal{F}(t,x) - \frac{1}{\sigma(t)}\sum_{i=1}^{r}\log(\delta(t) - g_i(t,x)) \tag{5.29}$$

where $x \in \tilde{D}$, $\tilde{D} := \{x \in \mathbb{R}^n : g_i(t,x) < \delta(t), i \in \{1,\ldots,r\}\}$ is an open set such that $D \subset \tilde{D}$. In the next lemma, we indicate the approximation error.

Lemma 5.2 (See [4]) *Under Assumptions 7 and 8* $\forall\ \lambda^* \in \mathfrak{T}^*(t)$ *and* $t \geq 0$, *the following inequality holds*

$$|\mathcal{F}(t,\hat{x}^*) - \mathcal{F}(t,x^*)| \leq \frac{r}{\sigma(t)} + \delta(t)\left(\inf_{\lambda^*\in\mathfrak{T}(t)^*}\|\lambda^*\|\right) \tag{5.30}$$

Remark 5.5 Lemma 5.2 says that if $\sigma(t)$ and $\delta(t)$ are chosen so that right-hand side of (5.30) converges to zero, then the approximate trajectory $\hat{x}^*(t)$ converges to $x^*(t)$.

To track the optimal solution $\hat{x}^*(t)$ within a predefined time, we aim to design a dynamics so that the gradient of $\tilde{\mathcal{L}}$ defined in (5.29) vanishes within a predefined time.Consider the following proposed dynamics in terms of Lagrangian (5.29) as:

$$\dot{x}(t) = \begin{cases} -(\nabla^2\tilde{\mathcal{L}})^{-1}\left(\vartheta\psi(t,x,\sigma,\delta) + \frac{\partial\nabla\tilde{\mathcal{L}}}{\partial\sigma}\dot{\sigma} + \frac{\partial\nabla\tilde{\mathcal{L}}}{\partial\delta}\dot{\delta} + \frac{\partial\nabla\tilde{\mathcal{L}}}{\partial t}\right) & \text{for } t_0 \leq t < t_f, \\ -(\nabla^2\tilde{\mathcal{L}})^{-1}\left(\vartheta\nabla\tilde{\mathcal{L}} + \frac{\partial\nabla\tilde{\mathcal{L}}}{\partial\sigma}\dot{\sigma} + \frac{\partial\nabla\tilde{\mathcal{L}}}{\partial\delta}\dot{\delta} + \frac{\partial\nabla\tilde{\mathcal{L}}}{\partial t}\right), & \text{for } t \geq t_f, \end{cases} \tag{5.31}$$

where $\vartheta \in \mathbb{R}$ and satisfies $\vartheta > 1$, $\psi(t,x,\sigma,\delta) = (\psi_1,\psi_2,\ldots,\psi_n) \in \mathbb{R}^{n\times 1}$, $\psi_i := \frac{(e^{\nabla\tilde{\mathcal{L}}x_i(t,x,\sigma,\delta)}-1)}{e^{\nabla\tilde{\mathcal{L}}x_i(t,x,\sigma,\delta)}(t_f-t)}$, for $1 \leq i \leq n$.

Theorem 5.4 ([3]) *Consider the dynamics* (5.31). *Assume that there exist a continuously differentiable function* $V : [t_0,\infty)\times D \to \mathbb{R}_{\geq 0}$ *satisfying* $\gamma_1(x) \leq V(t,x) \leq \gamma_2(x)$, $\forall x \in D\setminus\{0\}, \forall t \in [t_0,\infty)$, $\gamma_1(x)$ *and* $\gamma_2(x)$ *be continuous positive definite functions,* $V(t,\hat{x}^*) = 0$ *and for* $V \neq 0$: $\dot{V} \leq 0, \forall t \geq t_f$ *and* $\dot{V} \leq \frac{-\vartheta(e^V-1)}{e^V(t_f-t)}$, $\forall t \in [t_0,t_f)$ *for* (5.31). *This signifies that* $x(t)$ *converges to the optimal solution* $\hat{x}^*(t)$ *in a priori chosen time.*

Proof Let us select the Lyapunov functional candidate

$$V = \|\nabla\tilde{\mathcal{L}}(t,x,\sigma,\delta)\|^2, \tag{5.32}$$

The derivative of the V along the dynamics (5.31) for $t \in [t_0, t_f)$

$$
\begin{aligned}
\dot{V}(x) =& 2(\nabla\tilde{\mathcal{L}})^\top \left(\nabla^2\tilde{\mathcal{L}}\dot{x} + \frac{\partial\nabla\tilde{\mathcal{L}}}{\partial\sigma}\dot{\sigma} + \frac{\partial\nabla\tilde{\mathcal{L}}}{\partial\delta}\dot{\delta} + \frac{\partial\nabla\tilde{\mathcal{L}}}{\partial t}\right) \\
=& 2(\nabla\tilde{\mathcal{L}})^\top \Bigg(\nabla^2\tilde{\mathcal{L}}\Big(-(\nabla^2\tilde{\mathcal{L}})^{-1}\Big(\vartheta\psi + \frac{\partial\nabla\tilde{\mathcal{L}}}{\partial\sigma}\dot{\sigma} + \frac{\partial\nabla\tilde{\mathcal{L}}}{\partial\delta}\dot{\delta} + \frac{\partial\nabla\tilde{\mathcal{L}}}{\partial t}\Big)\Big) \\
&+ \frac{\partial\nabla\tilde{\mathcal{L}}}{\partial\sigma}\dot{\sigma} + \frac{\partial\nabla\tilde{\mathcal{L}}}{\partial\delta}\dot{\delta} + \frac{\partial\nabla\tilde{\mathcal{L}}}{\partial t}\Bigg) \qquad (5.33) \\
=& -2\vartheta(\nabla\tilde{\mathcal{L}})^\top\psi(t, x, \sigma, \delta) \qquad (5.34)
\end{aligned}
$$

Equation (5.34) is similar to (5.7) of convergence analysis performed in Theorem 5.1. Hence, further analysis can be performed in the same way and finally, we get a similar expression as (5.10) in Theorem 5.1,

$$
\dot{\hat{V}} \leq -\frac{\hat{\vartheta}(e^{\hat{V}} - 1)}{e^{\hat{V}}(t_f - t)} \qquad (5.35)
$$

where $\hat{\vartheta} = \frac{\vartheta}{n} > 1$. Thus, inequality (5.35) represents predefined-time stability, which means the trajectory of (5.31) converges to the optimal solution of $\mathcal{F}(t, x)$ in a priori chosen predefined-time.

Next, convergence analysis for system (5.31) for $t \geq t_f$ is similar to Theorem 5.1 and we get,

$$
\begin{aligned}
\dot{V}(x) =& 2(\nabla\tilde{\mathcal{L}})^\top \left(\nabla^2\tilde{\mathcal{L}}\dot{x} + \frac{\partial\nabla\tilde{\mathcal{L}}}{\partial\sigma}\dot{\sigma} + \frac{\partial\nabla\tilde{\mathcal{L}}}{\partial\delta}\dot{\delta} + \frac{\partial\nabla\tilde{\mathcal{L}}}{\partial t}\right) \\
=& -2\vartheta(\nabla\tilde{\mathcal{L}}(t, x, \sigma, \delta))^\top\nabla\tilde{\mathcal{L}}(t, x, \sigma, \delta) \\
=& -2\vartheta V
\end{aligned}
$$

So, for $t \geq t_f$, the equilibrium point of system (5.31) is asymptotically stable, therefore trajectories of (5.31) will remain optimal. Hence, $x(t) \to x^*(t)$ as $t \to t_f$.
□

5.5 Numerical Examples

This section provides three simulation examples to illustrate the predefined-time stability for the continuous-time dynamics proposed in this note. Euler discretization is used for MATLAB implementation with a constant time step size of Δt. Here, the convergence time (t_{ac}) in seconds is equal to $t_{ac}/\Delta t$ iterations.

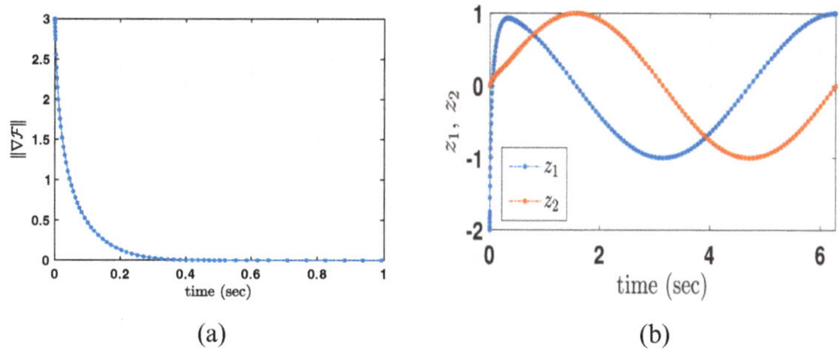

Fig. 5.1 Results using PTC-TVO dynamics (5.5) in solving example 1: (**a**) Comparison between $||\nabla\mathcal{F}||$ with PTC-TVO with $t_f = 0.6$ s and FxTC in [5], (**b**) Evolution of states using dynamics (5.5) for example 1 based on TVO problem with unconstrained case

5.5.1 Example 1

Minimize

$$\frac{1}{2}\begin{bmatrix} z_1 - \cos(t) \\ z_2 - \sin(t) \end{bmatrix}^\top \begin{bmatrix} 3 - 2\cos(2t) & 2\sin(2t) \\ 2\sin(2t) & 3 + 2\cos(2t) \end{bmatrix} \begin{bmatrix} z_1 - \cos(t) \\ z_2 - \sin(t) \end{bmatrix}$$

First, we implemented predefined-time convergent proposed dynamics (5.5) in MATLAB to solve the above TVO problem. The initial conditions are $z_0 = [-2 \ \ 0]^\top$. Figure 5.1a shows that the norm of the gradient of objective function vanishes in predefined-time 0.6 s and the comparison of the result with fixed-time convergent (FxTC) algorithm discussed in Hong et al. [5] is also shown. Figure 5.1b shows the evolution of states which starts tracking the optimal trajectory $z^*(t)$ in time $t_f = 0.6$ s, which is chosen a priori. In Fig. 5.2, we plot $\|\mathcal{F}(t, z)\|$ versus t for different sampling periods $\Delta t \in \{10^{-2}, 10^{-3}, 10^{-4}\}$. We notice that steady state error increases as we increase Δt. The log scale is used on the y axis such that the variation of $\|\nabla\mathcal{F}\|$ for values near zero is clearly shown.

Next, we show the comparison results with the prediction-correction method based on standard Newton flow dynamics presented in Fazlyab et al. [4] as: $\dot{z}(t) = -(\nabla^2\mathcal{F}(t, z))^{-1}\left(\vartheta\nabla\mathcal{F}(t, z) + \frac{\partial\nabla\mathcal{F}(t,z)}{\partial t}\right)$. It is shown in Fig. 5.3a that with the same initial condition $z_0 = [-2 \ 0]^\top$ and $\vartheta = 5$, $\|\nabla\mathcal{F}\|$ vanishes within predefined time $t_f = 0.3$ s using proposed dynamics (5.5) and using standard Newton flow dynamics discussed in Fazlyab et al. [4], $\|\nabla\mathcal{F}\|$ takes more than $t = 1.2$ s in reaching zero. In Fig. 5.3b, it shown that z_1, z_2 calculated using proposed dynamics (5.5) starts tracking optimal trajectory within predefined time $t_f = 0.3$ s while using standard Newton Flow dynamics, optimal trajectory tracking starts after $t = 1.2$ s.

Next, we solved the above example using a robust dynamics proposed in (5.11). We took initial conditions as $z_0 = [-2 \ 0]^\top$, $t_f = 0.6$ s, $\ell = 2$, $\vartheta = 5$. Figure 5.4a shows that the norm of the gradient of the objective function vanishes

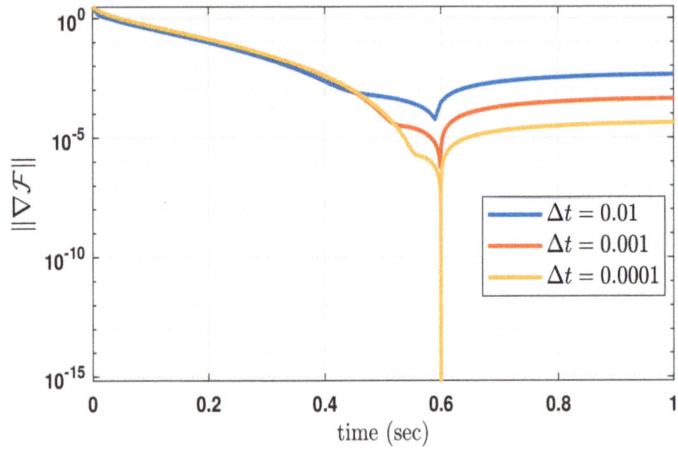

Fig. 5.2 Plot of $\|\nabla\mathcal{F}(t, z(t))\|$ on semilog scale, where $z(t)$ is the solution to the discretized version of dynamics (5.5) with time for various step sizes (Δt)

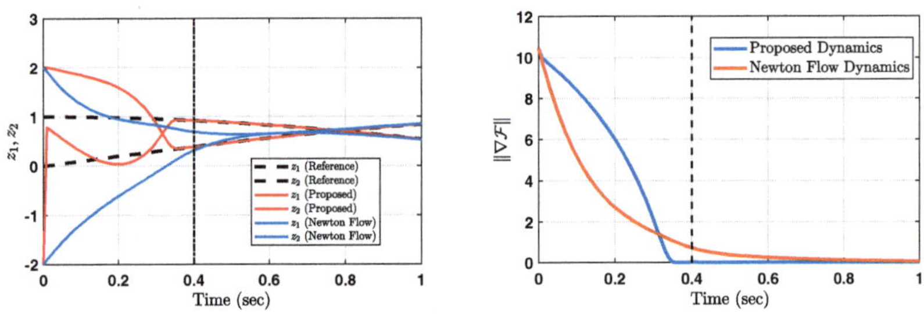

(a) Comparison of the evolution of states using dynamics (5.5) and Newton flow dynamics in Fazlyab et.al. (2017) [61] for example 1.

(b) Comparison between $||\nabla\mathcal{F}||$ with PTC-TVO with t_f = 0.4 sec and standard Newton flow dynamics in Fazlyab et.al. (2017) [61]

Fig. 5.3 Comparison of results using PTC-TVO dynamics (5.5) and standard Newton flow dynamics in Fazlyab et al. [4] in solving example 1. (**a**) Comparison of the evolution of states using dynamics (5.5) and Newton flow dynamics in Fazlyab et al. [4] for example 1. (**b**) Comparison between $||\nabla\mathcal{F}||$ with PTC-TVO with $t_f = 0.4$ s and standard Newton flow dynamics in Fazlyab et al. [4]

in predefined-time 0.6 s. In Fig. 5.4b, log scale is used on the y axis such that the variation of $\|\nabla\mathcal{F}\|$ for values near zero is clearly shown. Further, we show the comparison results from Newton flow robust dynamics for solving TVO problems: $\dot{z}(t) = -(\nabla^2\mathcal{F}(t, z))^{-1}\left(\vartheta\nabla\mathcal{F}(t, z) + \ell\frac{\nabla\mathcal{F}(t,z)}{\|\nabla\mathcal{F}(t,z)\|}\right)$. It is shown in Fig. 5.3a that with the same initial condition $z_0 = [-2\ 0]^\top$, $t_f = 0.6$ s, $\ell = 2$, $\vartheta = 5$, $\|\nabla\mathcal{F}\|$ vanishes within predefined time $t_f = 0.6$ s using proposed dynamics (5.11) and using standard robust Newton flow dynamics, $\|\nabla\mathcal{F}\|$ takes more than $t = 0.8$ s in reaching zero. In Fig. 5.3b, it shown that z_1, z_2 calculated using proposed dynamics (5.11) starts tracking optimal trajectory within predefined time $t_f = 0.6$ s while

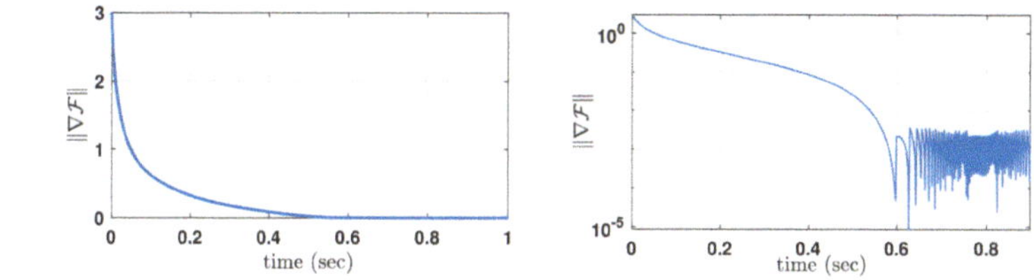

(a) Plot of $||\nabla\mathcal{F}||$ showing predefined-time convergence within $t_f = 0.6$ sec using dynamics (5.11) for Example 1 of unconstrained case.

(b) Plot of $||\nabla\mathcal{F}(t, z(t))||$ on semilog scale where $z(t)$ is the solution to the discretized version of (5.11) with sampling time $\Delta t = 0.001 sec$

Fig. 5.4 Results using PTC-TVO dynamics (5.11) in solving example 1. (**a**) Plot of $||\nabla\mathcal{F}||$ showing predefined-time convergence within $t_f = 0.6$ s using dynamics (5.11) for Example 1 of unconstrained case. (**b**) Plot of $||\nabla\mathcal{F}(t, z(t))||$ on semilog scale where $z(t)$ is the solution to the discretized version of (5.11) with sampling time $\Delta t = 0.001$ s

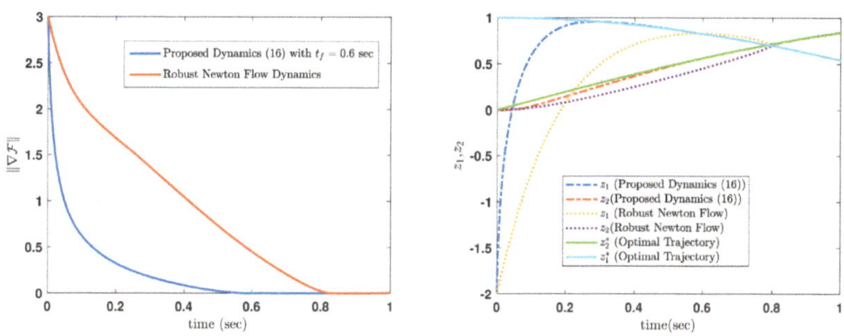

(a)

(b)

Fig. 5.5 Comparison of results using PTC-TVO dynamics (5.11) and standard robust Newton flow dynamics in solving example 1

using standard robust Newton Flow dynamics, optimal trajectory tracking starts after $t = 0.8$ s (Fig. 5.5).

5.5.2 Example 2

Consider the following quadratic TVO problem with ill-conditioned Hessian

$$\text{minimize} \quad (tz - \sin(t))^2 \tag{5.36}$$

Since, for this example, Hessian will be ill-conditioned, when t is small. Therefore, we apply the proposed Levenberg–Marquardt-like PTC-TVO algorithm (5.17) to

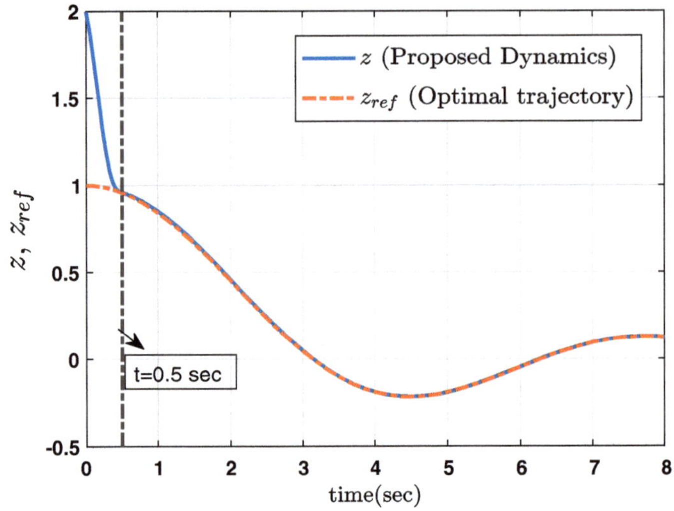

Fig. 5.6 Evolution of state using dynamics (5.17) for example (5.36) with ill-conditioned Hessian and it is shown that it starts tracking optimal trajectory within $t_f = 0.5$ s

find an optimal trajectory in a prior chosen time. It is important to choose $\wp$ properly. Here, we chose $\wp = e^{-2t}$. The initial condition is $z(0) = 2$, $t_f = 0.5$ s.

It is shown in Fig. 5.6 that the state trajectory obtained by using proposed dynamics (5.17) starts tracking the optimal trajectory within chosen $t_f \leq 0.5$ s.

5.5.3 *Example 3*

Consider the following quadratic TVO problem with inequality constraints:

$$\begin{aligned} &\text{minimize} \quad 0.5(z_1 + \sin(t))^2 + 1.5(z_2 + \cos(t))^2 \\ &\text{subject to} \quad z_2 - z_1 - \cos(t) \leq 0 \end{aligned} \tag{5.37}$$

Above example is an illustration of (5.22), so (5.29) is written as

$$\begin{aligned} \mathcal{L}(t, z, \sigma, \delta) &= 0.5(z_1 + \sin(t))^2 + 0.5(z_2 + \cos(t))^2 \\ &\quad - \frac{1}{\sigma(t)} \log(\delta(t) - (z_2 - z_1 - \cos(t))) \end{aligned} \tag{5.38}$$

where $\sigma(t) = 10 \exp(1000 * t)$, $\delta(t) = 2 \exp(-5t)$. We solved it using predefined-time convergent proposed dynamics (5.31) with the help of forward Euler's method with constant step size in MATLAB. The initial condition is $z_0 = [-2\ 0]^\top$. In

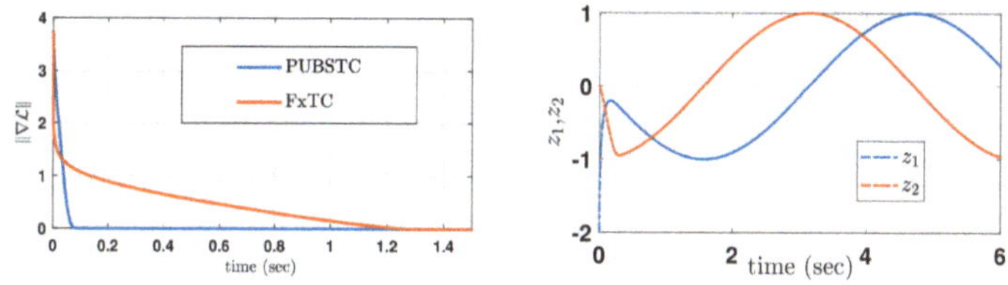

(a) Comparison between $||\nabla\mathcal{L}||$ with predefined-time convergence within t_f = 0.1 sec and FxTC.

(b) Evolution of states for example based on TVO problem with inequality constraints.

Fig. 5.7 Results using PTC-TVO dynamics (5.31) in solving example 2. (**a**) Comparison between $||\nabla\mathcal{L}||$ with predefined-time convergence within $t_f = 0.1$ s and FxTC. (**b**) Evolution of states for example based on TVO problem with inequality constraints

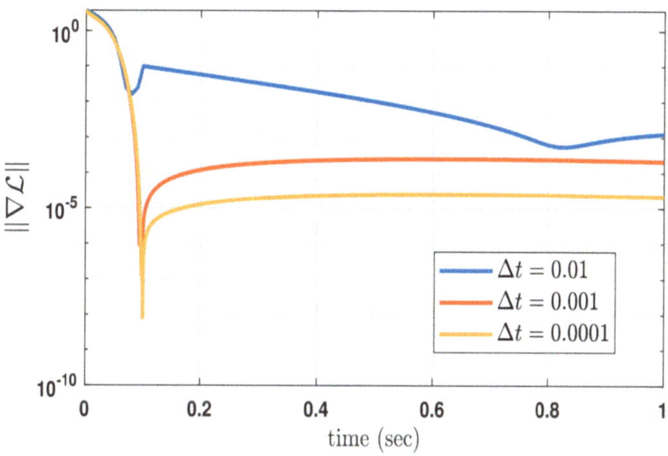

Fig. 5.8 Plot of $\|\nabla\mathcal{L}(t, z(t))\|$ on semilog scale, where $z(t)$ is the solution to the discretized version of dynamics (5.31) with time for various step sizes (Δt)

Fig. 5.7a, the norm of the gradient of $\nabla\mathcal{L}$ in (5.38) goes to zero within a priori chosen time of $t_f = 0.1$ s, which is the optimality condition in order to track the optimal trajectory and the comparison of the result with fixed-time convergent (FxTC) algorithm discussed in Hong et al. [5] is also shown. In Fig. 5.7b, the evolution of time-varying states is also shown which follow their optimal trajectory as $\|\nabla\mathcal{L}(t, z(t))\|$ vanishes within predefined-time $t_f = 0.1$ s. In Fig. 5.8, we plot $\|\nabla\mathcal{L}(t, z)\|$ versus t for different $\Delta(t) \in \{10^{-2}, 10^{-3}, 10^{-4}\}$. We notice that steady state error increases as we increase Δt.

5.6 Conclusion

In this chapter, predefined-time convergent dynamics are proposed and discussed to solve TVO problems. The proposed method yields a better convergence time compared to the methods available in the literature. Numerical examples show the effectiveness of proposed dynamics. Further, the proposed technique can also be extended for time-varying distributed optimization problems. In addition, one of the motivations behind the study of fast optimization methods is to generate signals for controllers at the same speed as the system evolves.

In the forthcoming chapter, our attention will be directed towards the application of the proposed approach in the present chapter in robot navigation with collision avoidance.

References

1. Karimi, H., Nutini, J., Schmidt, M.: Linear convergence of gradient and proximal-gradient methods under the Polyak-Łojasiewicz condition. In: Joint European Conference on Machine Learning and Knowledge Discovery in Databases, pp. 795–811. Springer (2016)
2. Boyd, S.P., Vandenberghe, L.: Convex Optimization. Cambridge University Press (2004)
3. Pandey, S., Prasun, P., Kamal, S., Singh, D., Ghosh, D., Olaru, S.: Optimal trajectory tracking using Newton and Levenberg-Marquardt-like algorithms for constrained time-varying optimization. Optimal Control Appl. Methods **46**(4), 1800–1813 (2025)
4. Fazlyab, M., Paternain, S., Preciado, V.M., Ribeiro, A.: Prediction-correction interior-point method for time-varying convex optimization. IEEE Trans. Autom. Control **63**(7), 1973–1986 (2017)
5. Hong, H., Yu, W., Jiang, G.-P., Wang, H.: Fixed-time algorithms for time-varying convex optimization. IEEE Trans. Circuits Syst. II Express Briefs **70**(2), 616–620 (2022)

Chapter 6
Application in Robot Navigation with Collision Avoidance

Abstract This chapter presents an application of predefined-time optimization dynamics to robot navigation with collision avoidance in cluttered environments. The navigation problem is formulated as a time-varying optimization problem, where a safe trajectory is computed using barrier-based constraints and power diagram concepts. A projected goal is obtained by solving the optimization problem in a predefined time, ensuring collision-free motion. A predefined-time controller is then designed to track this trajectory efficiently. Simulation results demonstrate accurate, fast, and safe navigation compared to existing methods.

6.1 Introduction

In this dedicated chapter, we consider the problem of navigation in sphere worlds, which involves designing a vector field over a compact, convex subset of Euclidean space punctured by Euclidean disks. The goal is to guide a Euclidean disk-shaped robot from almost all initial positions (excluding a set of measure zero) to a moving target, ensuring collision-free motion throughout the trajectory. We demonstrate the validity of our proposed predefined-time convergent dynamics to solve TVO problems with inequality constraints in robot navigation using simulation results.

The structure of this chapter is outlined as follows. In Sect. 6.2, we formulate the problem. In Sect. 6.3, we define the safe neighborhood of the robot by encoding collision. In Sect. 6.4, we formulate the robot navigation to a moving target as a TVO problem. Finally, in Sect. 6.5, the effectiveness of the proposed approach to solve the defined problem is validated through simulation results.

S. Kamal, S. Pandey, *A Continuous-Time Perspective on Accelerated Optimization and Control*, Studies in Systems, Decision and Control 693,
https://doi.org/10.1007/978-3-032-31645-5_6

6.2 Problem Formulation

In this chapter, we consider a disk-shaped robot centered at $z \in \mathcal{W}$ with radius $r \in \mathbb{R}_{\geq 0}$ working in a closed compact convex environment $\mathcal{W} \subset \mathbb{R}^n$. Let workspace $\mathcal{W}$ is populated with $m \in \mathbb{Z}_{\geq 0}$ disk shaped obstacles centered at $q_i \in \mathcal{W}$ with radius $\varrho_i > 0$ for $i \in \{1, 2, \ldots, m\}$. Hence, the free space $\mathcal{F}$ of the robot is defined as the set of configurations in the workspace in which the robot does not collide with any of the obstacles:

$$\mathcal{F} := \{z \in \mathcal{W} | \bar{\mathcal{B}}(z, r) \subseteq \mathcal{W} \backslash \bigcup_{i=1}^{m} \mathcal{B}(q_i, \varrho_i)\},$$

where, $\mathcal{B}(z, r)$ represents a n dimensional open ball centered at z with radius r, and $\bar{\mathcal{B}}(z, r)$ represents its closure. To ensure the connectivity of the free space $\mathcal{F}$, we assume that our disk-shaped robot is capable of navigating around any obstacle in any direction within the workspace $\mathcal{W}$.

Assumption 9 *Obstacles are separated from each other by clearance of at least*

$$\|q_i - q_j\| > \varrho_i + \varrho_j + 2r, \quad \forall i \neq j$$

and from the boundary $\partial\mathcal{W}$ of the workspace $\mathcal{W}$ as

$$\min_{p \in \partial\mathcal{W}} \|p - q_i\| > \varrho_i + 2r, \quad \forall i.$$

6.3 Encoding Collisions

In this section, we outline several key properties of power diagrams and explain how they are utilized to define a safe neighborhood for a robot.

6.3.1 Power Diagrams

The power diagram $P(q, \varrho) = \{P_1, P_2, ..., P_m\}$ of a convex environment $\mathcal{W} \in \mathbb{R}^n$, based on the set of generator disks centered at $q = (q_1, q_2, ..., q_m) \in \mathcal{W}^m$ with a vector of radii $\varrho = (\varrho_1, \varrho_2, ..., \varrho_m)$ is a partition of $\mathcal{W}$ such that every point $p \in \mathcal{W}$, is assigned to the closest generator based on the power distance $\|p - q_i\|^2 - \varrho_i^2$ as

$$P_i := \{p \in \mathcal{W} | \|p - q_i\|^2 - \varrho_i^2 \leq \|p - q_j\|^2 - \varrho_j^2, \ \forall j \neq i\},$$

A power diagram defines a convex cell decomposition of a convex environment, and the boundary ∂P_i of P_i is define by the boundary $\partial \mathcal{W}$ of the workspace $\mathcal{W}$ and the separating hyperplane H_{ij} between power cells P_i and P_j for some $j \neq i$ [1]. The separating hyperplane H_{ij} between any pair $i \neq j$ of power cells P_i and P_j is perpendicular to the line joining q_i and q_j and passes through the point $h_{ij} := \eta_{ij} p_i + (1 - \eta_{ij} p_j)$,

$$H_{ij} := \{p \in \mathbb{R}^n | (p - h_{ij})^\top (q_i - q_j) = 0\}$$

where, $\eta_{ij} := \frac{1}{2} - \frac{\varrho_i^2 - \varrho_j^2}{2\|q_i - q_j\|^2}$, and the perpendicular distance of qi to H_{ij} is given by

$$d(q_i, H_{ij}) := \min_{q \in H_{ij}} \|p - q_i\| = (1 - \eta_{ij})\|q_i - q_j\| = \varrho_i + \frac{(\|q_i - q_j\| - \varrho_i)^2 - \varrho_j^2}{2\|q_i - q_j\|}.$$

Objective Let z_c be the center of mass of the robot and our objective is for $z_c(t)$ to track a moving target $z_d(t) \in \mathcal{F}$ after a certain time $t > 0$ ensuring collision avoidance.

6.3.2 Safe Neighborhood of a Robot

Using the robot and obstacles as generator disks of a power diagram of $\mathcal{W}$, we define the local workspace, $\mathcal{LW}(z_c)$, of the robot as

$$\mathcal{LW}(z_c) := \{p \in \mathcal{W} | \|p - z_c\|^2 - r^2 < \|p - q_i\|^2 - \varrho_i^2, \ \forall i\}$$

To determine a collision-free neighborhood of the robot, we define the robot's local free space, $\mathcal{LF}(z_c)$, by eroding $\mathcal{LW}(z_c)$, removing the volume swept along its boundary, $\partial \mathcal{LW}(z_c)$, by the robot body radius, illustrated in Fig. 6.1.

Next, collision-free space around z_c is defined as [2]: $\mathcal{LF}(z_c) := \{p \in \mathcal{W} | c_i(z_c)^\top z - d_i(z_c) \leq 0, i = 1, ..., m\}$, where $c_i(z_c) = q_i - z_c$, $d_i(z_c) = (q_i - z_c)^\top \left(\eta_i q_i + (1 - \eta_i) z_c + r \frac{z_c - q_i}{\|z_c - q_i\|}\right)$, where, $\eta_i = \frac{1}{2} - \frac{\varrho_i^2 - r^2}{\|q_i - z_c\|^2}$.

Arslan and Koditschek [2], proposed the idea of a projected goal which constitutes a continuous computation of projection of $z_d(t)$ onto a safe neighborhood around z_c. Assuming the robot follows single integrator dynamics $\dot{z}_c = u(z_c)$, the controller proposed by Arslan and Koditschek [2] is $\dot{z}_c = -k(z_c - \hat{z})$, where $k > 0$ is gain and $\hat{z}$ is the orthogonal projection of z_d onto $\mathcal{LF}(z_c)$. The proposed controller works with the assumption that $\|z_i - z_j\| > r_i + r_j + 2r$.

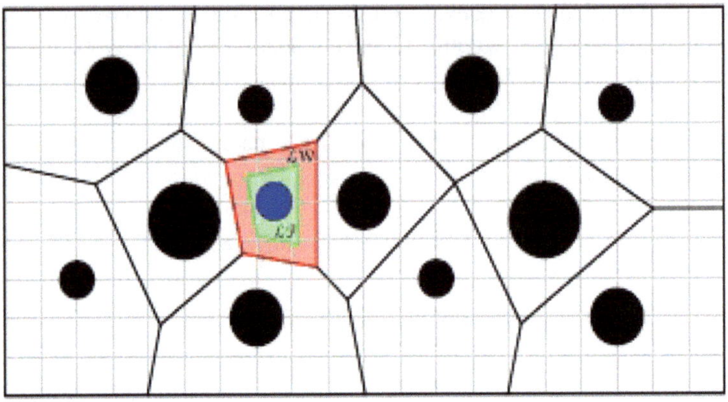

Fig. 6.1 Local workspace (red) and local free space (green) of a robot (blue) [2]

6.4 Formulation of Robot Navigation as a Time-Varying Optimization Problem

In this section, robot navigation problem discussed in previous section is reformulated as an optimization problem as follows:

$$\hat{z} := \underset{z\in\mathbb{R}^n}{\text{argmin}} \frac{1}{2}\|z - z_d(t)\|^2,$$

$$\text{subject to } \; c_i(z_c)^\top z - d_i(z_c) \leq 0, \quad i = 1, ..., m$$

Lagrangian for the above problem is written as (discussed in previous chapter)

$$\tilde{\mathcal{L}}(t, z, z_c) = \frac{1}{2}\|z - z_d(t)\|^2 - \frac{1}{\sigma(t)}\sum_{i=1}^{m} \log(d_i(z_c) - c_i(z_c)^\top z),$$

We solve the above problem using the proposed predefined-time convergent dynamics [4]

$$\dot{z}(t) = \begin{cases} -(\nabla^2\tilde{\mathcal{L}})^{-1}\left(\vartheta\psi(t, z, \sigma, \delta) + \frac{\partial\nabla\tilde{\mathcal{L}}}{\partial\sigma}\dot{\sigma} + \frac{\partial\nabla\tilde{\mathcal{L}}}{\partial\delta}\dot{\delta} + \frac{\partial\nabla\tilde{\mathcal{L}}}{\partial t}\right) & \text{for } t_0 \leq t < t_f, \\ -(\nabla^2\tilde{\mathcal{L}})^{-1}\left(\vartheta\nabla\tilde{\mathcal{L}} + \frac{\partial\nabla\tilde{\mathcal{L}}}{\partial\sigma}\dot{\sigma} + \frac{\partial\nabla\tilde{\mathcal{L}}}{\partial\delta}\dot{\delta} + \frac{\partial\nabla\tilde{\mathcal{L}}}{\partial t}\right), & \text{for } t \geq t_f, \end{cases} \tag{6.1}$$

where $\vartheta \in \mathbb{R}$ and satisfies $\vartheta > 1$, $\psi(t, z, \sigma, \delta) = (\psi_1, \psi_2, ..., \psi_n) \in \mathbb{R}^{n\times 1}$, $\psi_i = \frac{(e^{\nabla\tilde{\mathcal{L}}_{z_i}(t,z,\sigma,\delta)} - 1)}{e^{\nabla\tilde{\mathcal{L}}_{z_i}(t,z,\sigma,\delta)}(t_f - t)}$, for $1 \leq i \leq n$.

We propose a controller based on the Lemma 3.2 to navigate $z_c(t)$ to $\hat{z}(t)$ in a priori chosen time. Controller dynamics is:

$$u = \begin{cases} -k\frac{e^{(z_c(t)-\hat{z}(t))}-1}{e^{(z_c(t)-\hat{z}(t))}(t_f-t)}, & \text{for } t_0 \le t < t_f, \\ -k(z_c(t)-\hat{z}(t)), & \text{for } t \ge t_f, \end{cases} \tag{6.2}$$

where $k > 1$ is gain.

6.5 Simulation Results

In our simulation, we consider a moving target $z_d(t) = [x_d(t), y_d(t)]^\top \in \mathbb{R}^2$, following a circular trajectory with circumference of radius ($r_d = 15$), centered at origin, and moving periodically with $T = 2000$ s. We consider $m = 3$ obstacles in workspace $\mathcal{W}$, with centres $z_1 = [5, 5], z_2 = [15, 5], z_3 = [10, 15]$ with radii $r_1 = 1, r_2 = 2, r_3 = 1.5$ respectively. We consider a disk-shaped robot with a radius $r = 2$ with the initial point of its center of mass $z_c(0) = [2, 3]$. Next, we solve TVO problem to estimate the projected goal ($\hat{z}(t)$) $= [\hat{x}(t), \hat{y}(t)]$ using proposed continuous-time dynamics (6.1), where, $\widetilde{\mathcal{L}}(t, z, z_c) = \frac{1}{2}\|z - z_d(t)\|^2 - \frac{1}{\sigma(t)}\sum_{i=1}^{m}\log(d_i(z_c) - c_i(z_c)^\top z)$, here we took $\sigma(t) = e^{0.01t}$, $\vartheta = 50$, $t_f = 15$ s.

It is shown in Fig. 6.2 that $\|\nabla\mathcal{L}(t, z, z_c)\| \to 0$ as $t \to t_f$, which means we reach estimated optimal trajectory $\hat{z}(t)$ within $t_f = 15$ s.

It is shown in Fig. 6.3 that the trajectory of the center of mass of robot ($z_c(t)$) starts tracking $\hat{z}(t)$ within a predefined-time $t_f = 15$ s. Optimal trajectory $\hat{z}(t)$ provides safe navigation, avoiding obstacles in following the desired trajectory which is allowed to intersect obstacles as shown in Fig. 6.4. To track this moving projected target $\hat{z}(t)$ by the robot in a priori chosen time (t_f), we use controller (6.2), where for simulation, we chose gain $k = 5$.

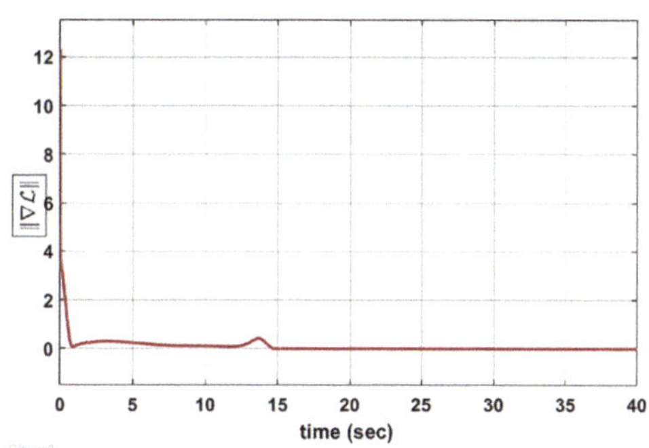

Fig. 6.2 Evolution of $||\nabla\mathcal{L}||$ with PUBST based convergence within $t_f = 15$ s using proposed continuous-time dynamics (5.31). It is shown that $\|\mathcal{L}\| \to 0$ as $t \to t_f$

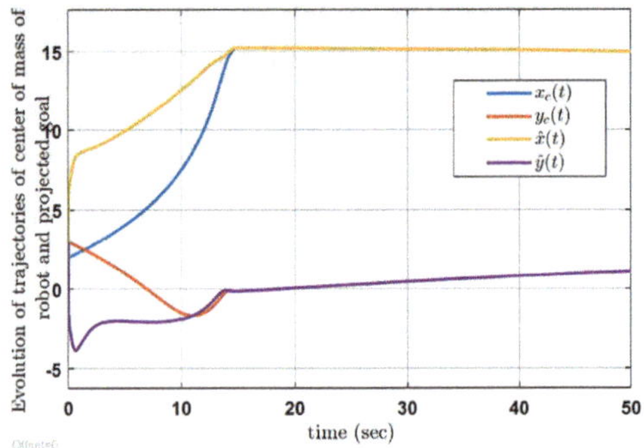

Fig. 6.3 Evolution of xy position of center of mass of robot $(x_c(t), y_c(t))$ and projected goal $(\hat{x}, \hat{y})$ evaluated solving TVO problem. It is shown that $z_c(t)$ starts tracking $\hat{z}(t)$ within $t_f = 15$ s

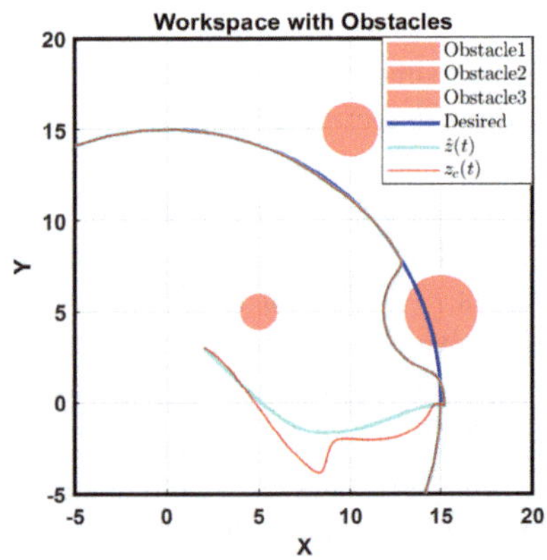

Fig. 6.4 Plot of workspace with three obstacles showing the evolution of desired trajectory $z_d(t)$, estimated projected goal $\hat{z}(t)$, and center of mass of robot $z_c(t)$

Figure 6.5 shows evolution of trajectories $z_c(t)$, $\hat{z}(t)$, and $z_d(t)$ over time. Figure 6.6 shows a phase portrait representing navigation of the robot. As we can see in Fig. 6.4, the robot successfully tracks the moving target while avoiding the circular obstacles.

In Fig. 6.7, we demonstrate that $\|\nabla\tilde{\mathcal{L}}\| \to 0$ within a predefined upper bound of settling time $t_f = 15$ s whereas dynamics proposed in Fazylab et al. [3] for solving TVO problems with inequality constraints takes more than $t = 40$ s to reach the estimated optimal trajectory. Additionally, we show the results obtained by plotting $\|\nabla\tilde{\mathcal{L}}\|$ on a semilog scale. Figure 6.8, illustrates that the steady-state error ($\|\nabla\tilde{\mathcal{L}}\| < 10^{-3}$ after $t_f > 15$ s) in the case of proposed dynamics is comparatively much smaller than the Newton flow dynamics. Figure 6.9 shows the plot of control inputs using dynamics (6.2) and the controller proposed by Arslan and Koditschek (2016) [2]: $\dot{z}_c = -k(z_c - \hat{z})$, where $k > 0$ is the gain.

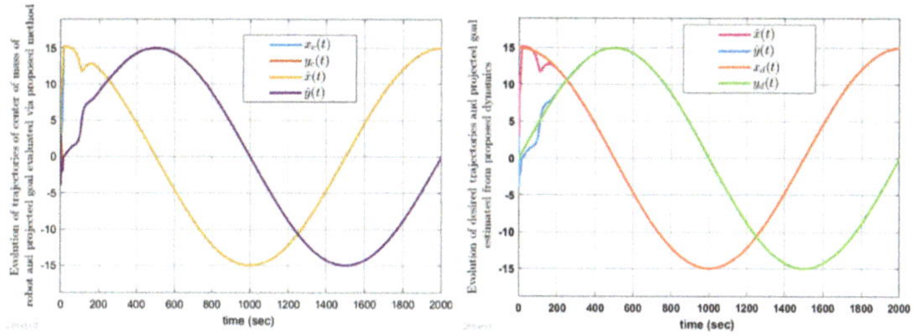

(a) Evolution of xy position of center of mass of robot $(x_c(t), y_c(t))$ and projected goal $(\hat{x}, \hat{y})$ evaluated solving TVO problem for complete time $T = 2000sec$.

(b) Evolution of xy position of the desired trajectory $(x_d(t), y_d(t))$ and projected goal $(\hat{x}, \hat{y})$ evaluated solving TVO problem for complete time $T = 2000sec$.

Fig. 6.5 Results using PTC-TVO dynamics (5.31) in solving TVO problem considered for computing projected goal required for safe robot navigation and plot of the desired trajectory which is moving periodically with $T = 2000$ s. (**a**) Evolution of xy position of center of mass of robot $(x_c(t), y_c(t))$ and projected goal $(\hat{x}, \hat{y})$ evaluated solving TVO problem for complete time $T = 2000$ s. (**b**) Evolution of xy position of the desired trajectory $(x_d(t), y_d(t))$ and projected goal $(\hat{x}, \hat{y})$ evaluated solving TVO problem for complete time $T = 2000$ s

Fig. 6.6 Phase-portrait: $x_c(t)$ opposite $y_c(t)$ showing that the center of mass of the robot follows a circular trajectory as desired over time

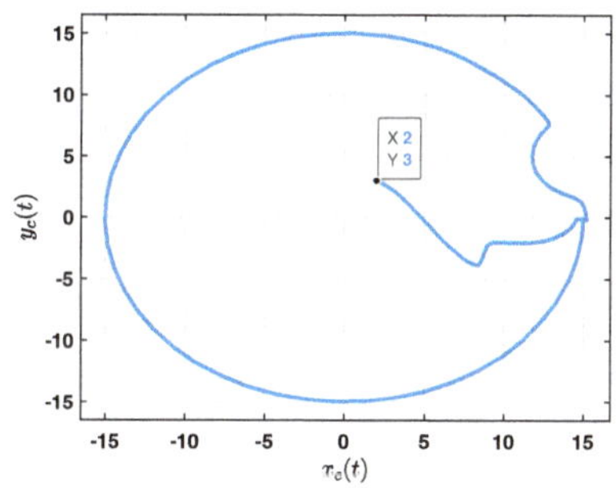

Fig. 6.7 Comparison of evolution of $\|\nabla\tilde{\mathcal{L}}\|$ with predefined-time convergence within $t_f = 15$ s using proposed dynamics (5.31) and Newton flow dynamics proposed in Fazlyab et al. [3]

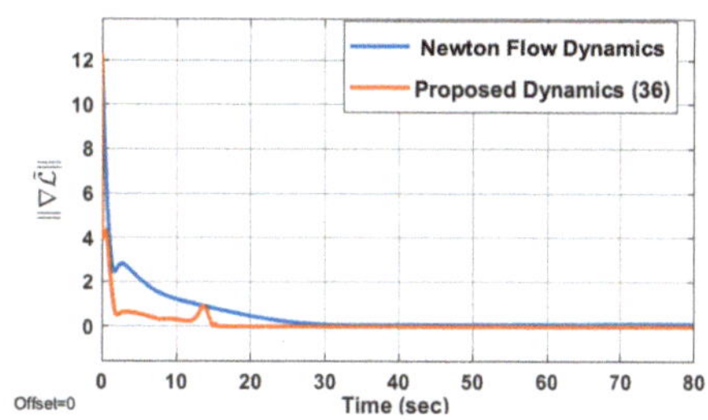

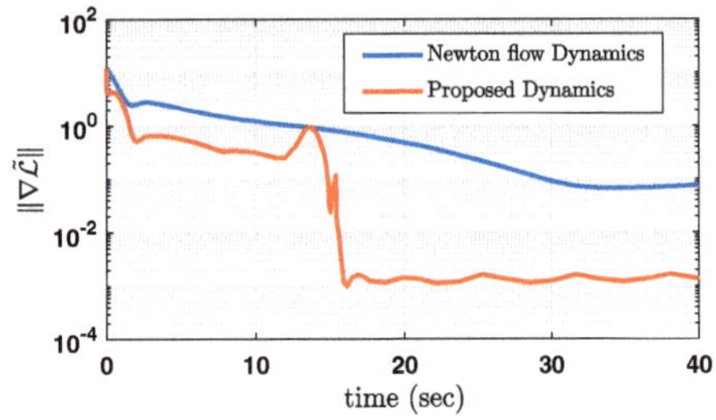

Fig. 6.8 Comparison of plot of $\|\nabla\tilde{\mathcal{L}}\|$ on semilog scale between our proposed dynamics to Newton flow dynamics discussed in Fazylab et al. [3] to show the accuracy of proposed dynamics

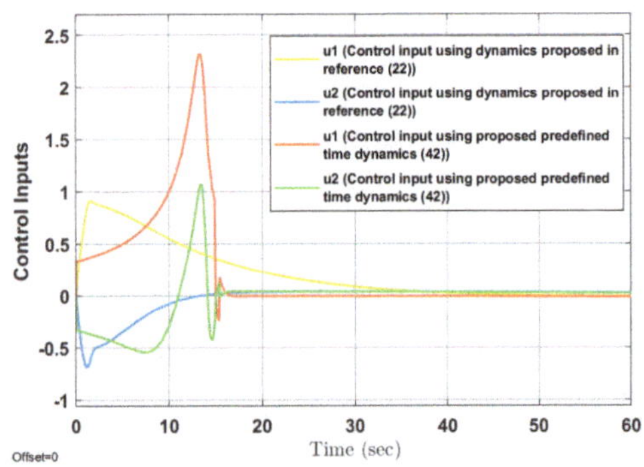

Fig. 6.9 Plot of control inputs using controller dynamics (6.2) and dynamics proposed in Arslan and Kod [2]

6.6 Conclusion

In conclusion, this chapter shows that the robot navigation to a moving target in the presence of obstacles can be formulated as a time-varying optimization problem. The formulated optimization problem is solved using the proposed predefined-time convergent dynamics, and a projected goal or optimal trajectory avoiding obstacles is found in the a priori chosen time. We apply a predefined-time controller to a robot so that it can also start tracking the found optimal projected goal in a user defined time.

References

1. Aurenhammer, F.: Power diagrams: properties, algorithms and applications. SIAM J. Comput. **16**(1), 78–96 (1987)
2. Arslan, O., Koditschek, D.E.: Exact robot navigation using power diagrams. In: 2016 IEEE International Conference on Robotics and Automation (ICRA), pp. 1–8. IEEE (2016)
3. Fazlyab, M., Paternain, S., Preciado, V.M., Ribeiro, A.: Prediction-correction interior-point method for time-varying convex optimization. IEEE Trans. Autom. Control **63**(7), 1973–1986 (2017)
4. Pandey, S., Prasun, P., Kamal, S., Singh, D., Ghosh, D., Olaru, S., Optimal trajectory tracking using Newton and Levenberg-Marquardt-like algorithms for constrained time-varying optimization. Optimal Control Appl. Methods **46**(4), 1800–1813 (2025)

Chapter 7
Feedback-Based Continuous-Time Optimization with Finite- and Fixed-Time Convergence

Abstract This chapter focuses on the design and analysis of feedback-based continuous-time optimization algorithms with guaranteed finite-time and fixed-time convergence properties. Motivated by the limitations of conventional asymptotic approaches and discrete-time methods, we adopt a dynamical systems perspective in which optimization problems are solved through appropriately designed continuous-time feedback laws. The framework integrates optimization objectives with nonlinear system dynamics, enabling autonomous operation without the need for external reference signals. Particular emphasis is placed on gradient-based feedback structures that ensure convergence to optimal solutions while accounting for the interaction between plant dynamics and control inputs. A central contribution of this chapter lies in the systematic development of finite-time and fixed-time stable optimization dynamics. Using Lyapunov-based analysis, sufficient conditions are established to guarantee convergence within a finite settling time, and further strengthened to ensure fixed-time convergence with bounds independent of initial conditions. The role of strong convexity and structural properties of the objective function is highlighted in achieving these guarantees. To rigorously analyze the coupled plant–controller system, time-scale separation techniques are employed, enabling decomposition into fast and slow subsystems and facilitating stability analysis of the overall closed-loop dynamics. The performance of the discussed framework is validated through its application to a nonlinear coupled-tank system, illustrating asymptotic, finite-time, and fixed-time convergence behaviors under different feedback designs.

7.1 Introduction

The integration of optimization and control has led to the development of feedback-based optimization frameworks, wherein optimization algorithms are embedded directly into the feedback loop of dynamical systems. Such approaches enable autonomous operation, eliminating the need for external reference signals and allowing the system to evolve naturally toward optimal steady-state behavior. This

S. Kamal, S. Pandey, *A Continuous-Time Perspective on Accelerated Optimization and Control*, Studies in Systems, Decision and Control 693,
https://doi.org/10.1007/978-3-032-31645-5_7

is particularly relevant in applications involving physical systems, such as power networks, robotics, and process control, where optimization and control must be performed simultaneously.

This chapter focuses on the systematic design and analysis of feedback-based CTO algorithms with finite-time and fixed-time convergence properties. We consider optimization problems coupled with nonlinear system dynamics and develop gradient-based and Newton-type feedback laws that ensure convergence to optimal solutions. The analysis is carried out using Lyapunov-based methods, which provide explicit conditions for stability and convergence rates.

A key aspect of the proposed framework is the use of time-scale separation techniques to analyze the interaction between plant dynamics and optimization dynamics. By decomposing the system into fast and slow subsystems, we establish rigorous convergence guarantees for the overall closed-loop system.

The contributions of this chapter are threefold. First, we develop a unified framework for feedback-based CTO with finite-time and fixed-time convergence guarantees. Second, we provide Lyapunov-based conditions that ensure stability and characterize convergence rates. Third, we validate the theoretical developments through application to a nonlinear coupled-tank system, demonstrating the effectiveness of the proposed methods.

The remainder of the chapter is organized as follows. We begin by formulating the optimization problem and reviewing preliminaries on continuous-time optimization and stability theory. We then develop feedback-based optimization laws and analyze their convergence properties under asymptotic, finite-time, and fixed-time regimes. Finally, we present simulation results and discuss potential extensions and future research directions.

7.2 Problem Formulation

In this section, we formalize the class of optimization problems considered in this chapter and introduce the dynamical system framework used for feedback-based continuous-time optimization. Consider a nonlinear dynamical system described by

$$\dot{x} = f(x, u), \quad x(0) = x_0,$$

where $x \in \mathbb{R}^n$ denotes the state vector, $u \in \mathbb{R}^m$ is the control input, and $f : \mathbb{R}^n \times \mathbb{R}^m \to \mathbb{R}^n$ is a locally Lipschitz continuous vector field. Under standard regularity assumptions, the system admits a unique solution for any initial condition and admissible input. We assume that the system possesses a steady-state mapping $h : \mathbb{R}^m \to \mathbb{R}^n$, satisfying

$$f(h(u), u) = 0, \quad \forall u \in \mathbb{R}^m,$$

which characterizes the equilibrium state corresponding to a constant input u. The mapping $h(\cdot)$ is assumed to be continuously differentiable and Lipschitz continuous.

7.3 Goal

The goal is to optimize a performance criterion defined by a continuously differentiable function $\phi : \mathbb{R}^n \times \mathbb{R}^m \to \mathbb{R}$. The associated constrained optimization problem is given by

$$\min_{x \in \mathbb{R}^n,\, u \in \mathbb{R}^m} \phi(x, u) \quad \text{subject to} \quad x = h(u). \tag{7.1}$$

By substituting the steady-state relation $x = h(u)$, the problem reduces to an unconstrained optimization problem in the input variable:

$$\min_{u \in \mathbb{R}^m} \hat{\phi}(u),$$

where $\hat{\phi}(u) := \phi(h(u), u)$. Throughout this chapter, we assume that ϕ (and hence $\hat{\phi}$) is convex and continuously differentiable. In certain cases, stronger assumptions such as strong convexity are imposed to guarantee the uniqueness of the optimal solution and improved convergence properties.

7.4 Feedback-Based Optimization Framework

Rather than solving the above optimization problem offline, our objective is to design a feedback-based control law that drives the system toward the optimal solution in real time. Specifically, we seek to construct a dynamical system of the form

$$\dot{x} = f(x, u), \quad \dot{u} = \psi(x, u),$$

where $\psi : \mathbb{R}^n \times \mathbb{R}^m \to \mathbb{R}^m$ is a feedback law designed such that the trajectories $(x(t), u(t))$ converge to an optimal pair (x^*, u^*). The equilibrium point (x^*, u^*) is said to be optimal if

$$x^* = h(u^*), \quad \nabla\hat{\phi}(u^*) = 0.$$

In Fig. 7.1, interconnection between the plant dynamics and feedback optimizer is shown.

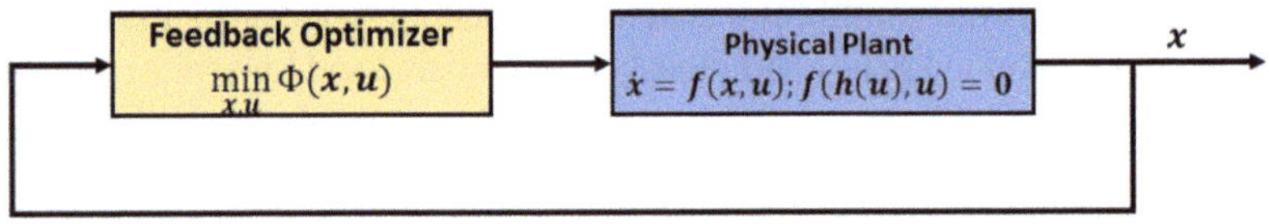

Fig. 7.1 Feedback-based continuous-time optimization

7.4.1 Separation of Time Scales in Autonomous Optimization

Proposition 7.1 ([1]) *Consider the non-linear dynamics* (7.1). *Assume that* $\exists$ *a differentiable,* ℓ*-Lipschitz map* $h : \mathbb{R}^p \to \mathbb{R}^m$ *s.t.:* $f(h(u), u) = 0 \quad \forall u \in \mathbb{R}^p$ *and for all* $x_0 \in \mathbb{R}^n$*, constant input* $\hat{u}$*, there exists* $\tau > 0$*,* $K > 0$ *such that:*

$$\|x(t) - h(\hat{u})\| \leq K e^{-\tau t}$$

Then, there exists a Lyapunov function $W(x, u)$ *and constants* $\alpha, \beta, \gamma, \zeta > 0$ *such that:*

$$\alpha \|x - h(u)\|^2 \leq W(x, u) \leq \beta \|x - h(u)\|^2$$
$$\dot{W}(x, u) \leq -\gamma \|x - h(u)\|^2$$
$$\|\nabla_u W(x, u)\| \leq \zeta \|x - h(u)\|$$

7.4.2 Gradient-Based Feedback Law

A gradient-based feedback control law is designed, leading to the following closed-loop system dynamics:

$$\begin{aligned} \dot{x} &= f(x, u), \\ \dot{u} &= -Q(u) H(u)^\top \nabla \Phi(x, u), \end{aligned} \tag{7.2}$$

where $Q : \mathbb{R}^p \to \mathbb{R}^{m \times m}$ is a Lipschitz continuous positive definite metric. The matrix $H(u)^\top$ is defined using the chain rule as

$$H(u)^\top := \left[\nabla h(u)^\top \; \mathbb{I}_p \right],$$

which captures the sensitivity of the constraint mapping $h(u)$ along with the identity component for the joint optimization structure.

7.4.3 Stability Analysis

Lemma 7.1 ([1]) *Every minimizer $(x^\star, u^\star)$ of* (7.1) *corresponds to an equilibrium point of* (7.2)*. Conversely, each equilibrium point of* (7.2) *is a critical point of* (7.1).

Lemma 7.2 *Consider the dynamical system*

$$\dot{x} = f(x, u), \quad \dot{u} = -Q(u)\, H(u)^\top \nabla\Phi(x, u),$$

under the conditions of Proposition 7.1. Suppose that the function $\Phi(h(u), u)$ admits compact sublevel sets and satisfies

$$\sup_{u \in \mathbb{R}^p} \|Q(u)\| < \frac{\gamma}{\zeta L}.$$

Then, all solutions are complete and converge to the set of first-order optimal points of the problem

$$\min\{\Phi(x, u) \mid x = h(u)\}.$$

Moreover, strict local minimizers correspond to stable equilibria, while asymptotically stable equilibria are necessarily strict local minimizers.

Proof Define a composite Lyapunov function:

$$\Psi(x, u) = (1 - \Delta)\tilde{\Phi}(u) + \Delta W(x, u), \quad 0 < \Delta < 1,$$

with $\tilde{\Phi}(u) = \Phi(h(u), u)$, and $W(x, u)$ a Lyapunov function satisfying:

$$\dot{W}(x, u) \le -\gamma \|x - h(u)\|^2, \quad \|\nabla_u W(x, u)\| \le \zeta \|x - h(u)\|.$$

The derivative of Ψ along trajectories:

$$\dot{\Psi}(x, u) \le \begin{bmatrix} \|g\|_{Q(u)} \\ \|x - h(u)\| \end{bmatrix}^\top \Lambda \begin{bmatrix} \|g\|_{Q(u)} \\ \|x - h(u)\| \end{bmatrix}$$

with $g(x, u) = -H(u)^\top \nabla\Phi(x, u)$ and matrix:

$$\Lambda = \begin{bmatrix} -(1 - \Delta) & \frac{1}{2}\kappa L(1 - \Delta) + \frac{1}{2}\kappa\zeta\Delta \\ \frac{1}{2}\kappa L(1 - \Delta) + \frac{1}{2}\kappa\zeta\Delta & -\gamma\Delta \end{bmatrix}$$

If $\Lambda \prec 0$, then $\dot{\Psi} \le 0$, and by LaSalle's theorem, the system converges to the optimal set. □

7.4.4 Finite-Time Controller Design

Consider the reduced optimization problem

$$\min_{u \in \mathbb{R}^m} \hat{\phi}(u),$$

where $\hat{\phi} \in C^2$ is assumed to be μ-strongly convex, ensuring the existence of a unique minimizer u^*. We propose the following nonlinear feedback law:

$$\dot{u} = -\epsilon \frac{\nabla \hat{\phi}(u)}{\|\nabla \hat{\phi}(u)\|^{\frac{\beta-2}{\beta-1}}}, \quad \beta > 2, \ \epsilon > 0,$$

in conjunction with the plant dynamics $\dot{x} = f(x, u)$.The non-Lipschitz nature of the feedback law in the vicinity of the equilibrium plays a crucial role in achieving finite-time convergence.

Assumption 10 *The plant dynamics satisfy*

$$(x - h^*)^T f(x, u) \leq -\kappa ||x - h^*||^p$$

where $\kappa > 0, 0 < p < 2$ *and* $||x - h^*||$ *represents the boundary layer error with* h^* *being the critical value.*

Theorem 7.1 (Finite-Time Convergence [2]) *Consider the closed-loop system*

$$\dot{x} = f(x, u), \quad \dot{u} = -\epsilon \frac{\nabla \hat{\phi}(u)}{\|\nabla \hat{\phi}(u)\|^{\frac{\beta-2}{\beta-1}}}, \tag{7.3}$$

where $\beta > 2$ *and* $\epsilon > 0$, *and suppose that the plant dynamics satisfy Assumption 10. Let* $V_c : \mathbb{R}^n \times \mathbb{R}^m \mapsto \mathbb{R}_+$ *be a composite Lyapunov function for the closed-loop system. If the perturbed parameter* ϵ *satisfied* $\epsilon \in \left(\frac{q(0.5)^\zeta \lambda^{\zeta-1}}{\mu}, \frac{q(0.5)^\zeta (1-\lambda)^{\zeta-1}}{\mu}\right)$ *where* $\zeta = \frac{\beta}{2(\beta-1)}$, $0 < \lambda < 1, q > 0$ *and* $\beta > 2$, *then equilibrium point of* (7.3) *is finite-time stable with a settling time* $T_s \leq \frac{1}{q(1-\zeta)} V_c(0)^{(1-\zeta)}$.

Proof The composite Lyapunov function V_c associated with the system in (7.3) is defined as a convex combination of the Lyapunov candidate functions corresponding to the plant and the controller, given by

$$V_c = \lambda V_p + (1 - \lambda) V_u,$$

where $\lambda \in (0, 1)$.

The Lyapunov candidate corresponding to the plant dynamics is selected as

$$V_p = \frac{1}{2}\left(\|x - h^*\|^{2q} + \|\nabla\hat{\phi}(u)\|^2\right), \quad q > 0,$$

whereas the Lyapunov candidate for the controller is chosen as

$$V_u = \frac{1}{2}\|\nabla\hat{\phi}(u)\|^2.$$

Evaluating the time derivative of V_c along the trajectories of (7.3) results in

$$\dot{V}_c = \lambda\dot{V}_p + (1-\lambda)\dot{V}_u.$$

Considering the first term, we have

$$\begin{aligned}\lambda\dot{V}_p &= \lambda\left(\nabla_x V_p\dot{x} + \nabla_u V_p\dot{u}\right)\\ &= \lambda\Big(q\|x - h^*\|^{2q-2}(x-h^*)^\top f(x,u) - \epsilon\nabla^2\hat{\phi}(u)\,\|\nabla\hat{\phi}(u)\|^{\frac{\beta}{\beta-1}}\Big). \qquad (7.4)\end{aligned}$$

By selecting $\kappa = 0.5^\zeta\lambda^{\zeta-1}$, enforcing the relation $2q - 2 + p = 2q\zeta$, and defining

$$\zeta = \frac{\beta}{2(\beta-1)},$$

together with the μ-strong convexity condition $\nabla^2\hat{\phi}(u) \succeq \mu I$, it follows that

$$\lambda\dot{V}_p \leq -\lambda q\kappa\|x - h^*\|^{2q\zeta} - \epsilon\lambda\mu\|\nabla\hat{\phi}(u)\|^{2\zeta}.$$

Further, by imposing the condition $\epsilon \leq \frac{q(0.5)^\zeta\lambda^{\zeta-1}}{\mu}$, we obtain $\lambda\dot{V}_p \leq -q\lambda^\zeta V_p^\zeta$. Next, consider the controller term:

$$\begin{aligned}(1-\lambda)\dot{V}_u &= (1-\lambda)\nabla\hat{\phi}(u)^\top\dot{u}\\ &\leq -\mu\epsilon(1-\lambda)\|\nabla\hat{\phi}(u)\|^{2\zeta}. \qquad (7.5)\end{aligned}$$

By imposing the condition $\epsilon \leq \frac{q(0.5)^\zeta(1-\lambda)^{\zeta-1}}{\mu}$, it follows that $(1-\lambda)\dot{V}_u \leq -q(1-\lambda)^\zeta V_u^\zeta$.

Combining the above inequalities, we obtain

$$\dot{V}_c \leq -q\lambda^\zeta V_p^\zeta - q(1-\lambda)^\zeta V_u^\zeta.$$

Using standard inequalities (e.g., Minkowski-type bounds), this can be expressed as

$$\dot{V}_c \leq -q V_c^{\zeta}.$$

Since $\zeta \in (0, 1)$, the above inequality guarantees finite-time stability. In particular, the settling time satisfies $T_s \leq \frac{1}{q(1-\zeta)} V_c(0)^{1-\zeta}$. Hence, the optimal equilibrium of (7.3) is finite-time stable. □

Remark 7.1 The essential feature enabling finite-time convergence is the non-Lipschitz scaling of the gradient term, which induces a stronger decay rate of the Lyapunov function near the equilibrium compared to standard gradient flows.

7.5 Fixed-Time Feedback-Based Optimization

In this section, we extend the previous framework to achieve fixed-time convergence, i.e., convergence within a uniform time bound independent of initial conditions. This is accomplished by introducing a feedback structure that combines multiple nonlinear terms with distinct homogeneity properties.

7.5.1 *Fixed-Time Controller Design*

Consider the reduced optimization problem

$$\min_{u \in \mathbb{R}^m} \hat{\phi}(u),$$

where $\hat{\phi} \in C^2$ is μ-strongly convex with unique minimizer u^*. We propose the following feedback law with dual exponents:

$$\dot{u} = -\epsilon_1 \frac{\nabla \hat{\phi}(u)}{\|\nabla \hat{\phi}(u)\|^{\frac{\beta_1 - 2}{\beta_1 - 1}}} - \epsilon_2 \frac{\nabla \hat{\phi}(u)}{\|\nabla \hat{\phi}(u)\|^{\frac{\beta_2 - 2}{\beta_2 - 1}}},$$

where $\epsilon_1, \epsilon_2 > 0$, $\beta_1 > 2$, and $1 < \beta_2 < 2$. The plant dynamics is $\dot{x} = f(x, u)$.

The first term dominates far from the equilibrium, while the second term dominates near the equilibrium, ensuring convergence in a uniform time independent of initial conditions.

Assumption 11 *The system satisfies the following conditions:*

$$(x - h^*)^T f(x, u) \leq -\kappa_1 ||x - h^*||^{p_1} - \kappa_2 ||x - h^*||^{p_2}$$

where $\kappa_1, \kappa_2 > 0, 0 < p_1 < 2$, *and* $p_2 > 2$ *with* h^* *being the critical value.*

Theorem 7.2 ([2] Fixed-Time Convergence) *Consider the closed-loop system*

$$\dot{x} = f(x, u), \quad \dot{u} = -\epsilon_1 \frac{\nabla\hat{\phi}(u)}{\|\nabla\hat{\phi}(u)\|^{\frac{\beta_1 - 2}{\beta_1 - 1}}} - \epsilon_2 \frac{\nabla\hat{\phi}(u)}{\|\nabla\hat{\phi}(u)\|^{\frac{\beta_2 - 2}{\beta_2 - 1}}}, \tag{7.6}$$

with $\beta_1 > 2$, $1 < \beta_2 < 2$, *and* $\epsilon_1, \epsilon_2 > 0$. *Suppose that the plant satisfies assumption 11. Consider the composite Lyapunov function* V_c, *defined as a convex combination of the plant and controller Lyapunov candidates,* V_p *and* V_u, *respectively. If the perturbed parameter* ϵ_i *satisfies the condition* $\epsilon_i \in \left(\frac{0.5^{\gamma_i} q \lambda^{\gamma_i - 1}}{2\mu_f^{\frac{\alpha_i}{2}}}, \frac{0.5^{\gamma_i} q (1-\lambda)^{\gamma_i - 1}}{2\mu_f^{\frac{\alpha_i}{2}}} \right)$ *where* $i = 1, 2$, $\lambda \in (0, 1)$, $\alpha_i = \frac{\beta_i}{\beta_i - 1}$, *and* $\gamma_i = \frac{2+\alpha_i}{4}$, *then the equilibrium of* (7.6) *is FxTS with a settling time* $T_s \leq \frac{1}{q(1-\gamma_1)} + \frac{1}{q(1-\gamma_2)}$.

Proof A composite Lyapunov function V_c is formulated as a convex combination of the Lyapunov candidates corresponding to the plant and the controller:

$$V_c = \lambda V_p + (1 - \lambda) V_u,$$

where $\lambda \in (0, 1)$. Differentiating along the system trajectories yields

$$\dot{V}_c = \lambda \dot{V}_p + (1 - \lambda) \dot{V}_u.$$

The Lyapunov candidate corresponding to the plant dynamics is selected as

$$V_p = \frac{1}{2}\|x - h^*\|^{2q} + \frac{1}{2}\left(\hat{\phi}(u) - \hat{\phi}^*\right)^2, \quad q > 0,$$

and for the controller,

$$V_u = \frac{1}{2}\left(\hat{\phi}(u) - \hat{\phi}^*\right)^2.$$

Examining the first term, we obtain

$$\begin{aligned} \lambda \dot{V}_p &= \lambda q \|x - h^*\|^{2q-2}(x - h^*)^\top f(x, u) \\ &\quad - \lambda\left(\epsilon_1(\hat{\phi} - \hat{\phi}^*)\|\nabla\hat{\phi}\|^{\frac{\beta_1}{\beta_1 - 1}} + \epsilon_2(\hat{\phi} - \hat{\phi}^*)\|\nabla\hat{\phi}\|^{\frac{\beta_2}{\beta_2 - 1}}\right). \end{aligned} \tag{7.7}$$

Let $\alpha_i = \frac{\beta_i}{\beta_i-1}, \quad \gamma_i = \frac{2+\alpha_i}{4}, \quad i = 1, 2$. Applying the Lojasiewicz inequality along with the conditions in Assumption 11, it follows that

$$\begin{aligned}\lambda \dot{V}_p \leq & -\kappa_1 \lambda q \|x - h^*\|^{2q\gamma_1} - \epsilon_1 \lambda (2\mu_f)^{\frac{\alpha_1}{2}} (\hat{\phi} - \hat{\phi}^*)^{2\gamma_1} \\ & - \kappa_2 \lambda q \|x - h^*\|^{2q\gamma_2} - \epsilon_2 \lambda (2\mu_f)^{\frac{\alpha_2}{2}} (\hat{\phi} - \hat{\phi}^*)^{2\gamma_2}. \end{aligned} \tag{7.8}$$

By choosing $\kappa_i = (0.5)^{\gamma_i} \lambda^{\gamma_i - 1}, \quad \epsilon_i = \frac{(0.5)^{\gamma_i} q \lambda^{\gamma_i - 1}}{\mu_f^{\frac{\alpha_i}{2}}}$, we obtain

$$\lambda \dot{V}_p \leq -q \left(\lambda^{\gamma_1} V_p^{\gamma_1} + \lambda^{\gamma_2} V_p^{\gamma_2} \right).$$

Next, consider the second term:

$$\begin{aligned}(1 - \lambda) \dot{V}_u = & -\epsilon_1 (1 - \lambda) (\hat{\phi} - \hat{\phi}^*) \|\nabla \hat{\phi}\|^{\frac{\beta_1}{\beta_1 - 1}} \\ & - \epsilon_2 (1 - \lambda) (\hat{\phi} - \hat{\phi}^*) \|\nabla \hat{\phi}\|^{\frac{\beta_2}{\beta_2 - 1}}. \end{aligned} \tag{7.9}$$

Using strong convexity, this can be bounded as

$$\begin{aligned}(1 - \lambda) \dot{V}_u \leq & -2(1 - \lambda) \epsilon_1 \mu_f^{\frac{\alpha_1}{2}} (\hat{\phi} - \hat{\phi}^*)^{2\gamma_1} \\ & - 2(1 - \lambda) \epsilon_2 \mu_f^{\frac{\alpha_2}{2}} (\hat{\phi} - \hat{\phi}^*)^{2\gamma_2}. \end{aligned} \tag{7.10}$$

Consequently, the perturbed parameters must satisfy the following condition:

$$\epsilon_i = \frac{(0.5)^{\gamma_i} q (1 - \lambda)^{\gamma_i - 1}}{\mu_f^{\frac{\alpha_i}{2}}}.$$

Combining the bounds for $\dot{V}_p$ and $\dot{V}_u$, we obtain

$$\dot{V}_c \leq -q \left((\lambda V_p + (1 - \lambda) V_u)^{\gamma_1} + (\lambda V_p + (1 - \lambda) V_u)^{\gamma_2} \right),$$

or equivalently, $\dot{V}_c \leq -q \left(V_c^{\gamma_1} + V_c^{\gamma_2} \right)$.

Thus, under the admissible range of the perturbed parameters

$$\epsilon_i \in \left(\frac{(0.5)^{\gamma_i} q \lambda^{\gamma_i - 1}}{\mu_f^{\frac{\alpha_i}{2}}}, \frac{(0.5)^{\gamma_i} q (1 - \lambda)^{\gamma_i - 1}}{\mu_f^{\frac{\alpha_i}{2}}} \right),$$

the system is FxTS at its optimal equilibrium point. Moreover, the settling time admits a uniform bound

$$T_s \leq \left| \frac{1}{q(1-\gamma_1)} + \frac{1}{q(\gamma_2 - 1)} \right|.$$

□

Remark 7.2 The use of dual exponents is essential for fixed-time convergence: the term with exponent less than one ensures rapid convergence near the equilibrium, while the term with exponent greater than one guarantees bounded convergence time for large initial conditions.

7.6 Simulation Results

7.6.1 Finite-Time Feedback Optimization

To illustrate the effectiveness of the proposed finite-time feedback optimization scheme, we consider a nonlinear coupled-tank system governed by the dynamics

$$\dot{x} = -\frac{A_{01}}{A_{t1}}\sqrt{2g\,x} + \frac{K_p}{A_{t1}}u, \tag{7.11}$$

where $x \in \mathbb{R}_+$ represents the liquid level in the tank, and $u \in \mathbb{R}_+$ denotes the control input (inflow rate). The parameters $A_{01} = 0.17812\,\text{cm}^2$ and $A_{t1} = 15.5179\,\text{cm}^2$ correspond to the cross-sectional areas of the outlet orifice and the tank, respectively. The constant $K_p = 3.3$ is a proportional gain, and $g = 981\,\text{cm/s}^2$ denotes gravitational acceleration.

The initial conditions for the simulation are chosen as

$$x(0) = 5, \quad u(0) = 3.$$

The objective is to regulate the system to a desired operating point by minimizing the cost function

$$\phi(x, u) = (x - 1)^2.$$

Accordingly, the steady-state optimization problem is formulated as

$$\min_{x,u} \ \phi(x, u) \quad \text{subject to} \quad x = \frac{K_p^2}{2gA_{01}^2}u^2. \tag{7.12}$$

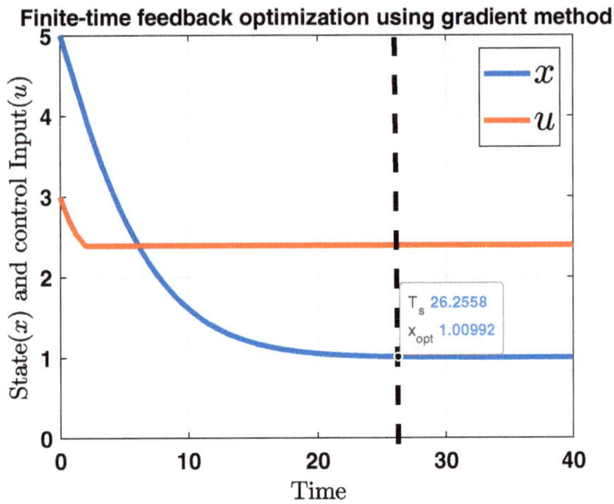

Fig. 7.2 The objective function yields the optimal operating point $(x, u) = (1, 0)$, demonstrating that the closed-loop system achieves the optimal steady-state under finite-time stability (FTS)

Substituting the steady-state relation yields the reduced objective function $\hat{\phi}(u)$, which is minimized using the proposed finite-time feedback controller. The control law is given by

$$\dot{u} = -0.4986 \frac{\nabla\hat{\phi}(u)}{\|\nabla\hat{\phi}(u)\|^{0.6}},$$

corresponding to the general form in (7.3), with parameters $\beta = 3.5$, $\lambda = 0.4$, $q = 1.37$, and $\mu = 2$.

The simulation results shown in Fig. 7.2 indicate that the proposed finite-time optimizer successfully drives the system states to the optimal operating point. In particular, the closed-loop system achieves convergence within a finite settling time bounded by $T_s \leq 26$ s, as illustrated in Fig. 7.2.

7.6.2 Fixed-Time Feedback Optimization

We next evaluate the performance of the discussed fixed-time feedback optimization scheme on the same coupled-tank system described in (7.11) and the optimization problem (7.12).

To achieve fixed-time convergence, the control input is updated using a dual-exponent feedback law of the form

$$\dot{u} = -0.4586 \frac{\nabla\hat{\phi}(u)}{\|\nabla\hat{\phi}(u)\|^{0.6}} - 0.1192 \frac{\nabla\hat{\phi}(u)}{\|\nabla\hat{\phi}(u)\|^{-0.25}},$$

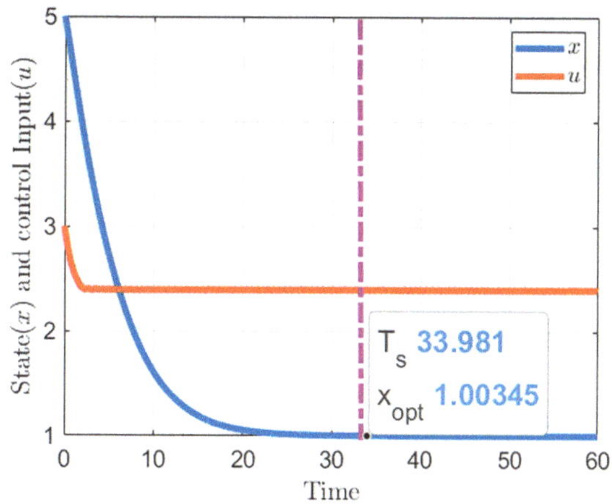

Fig. 7.3 Fixed-time convergence of the coupled-tank system under the proposed feedback law

which corresponds to the general fixed-time structure developed in the previous section.

The controller parameters are selected as $\beta_1 = 3.5, \beta_2 = 1.8, q = 2.5, \mu_f = 4, \lambda = 0.4$.

The first term in the control law dominates when the system state is far from the equilibrium, ensuring rapid global convergence, while the second term becomes dominant in the vicinity of the equilibrium, enforcing fast local convergence. This combination guarantees a uniform UBST, independent of the initial conditions.

Simulation results presented in Fig. 7.3 demonstrate that the proposed fixed-time controller successfully regulates the system to the optimal operating point. In contrast to the finite-time case, the convergence time remains bounded uniformly across different initial conditions, illustrating the FxTS property.

In particular, the closed-loop system achieves convergence within a prescribed bound $T_s \leq 33$ s, as depicted in Fig. 7.3. These results confirm that the dual-exponent feedback design significantly enhances convergence performance, providing both fast transient response and robustness with respect to initial conditions.

7.7 Conclusion

This chapter developed a feedback-based continuous-time optimization framework with guaranteed FTC and FxTC. Starting from gradient-based dynamics, we designed nonlinear feedback laws that ensure accelerated convergence to the optimal solution. Using Lyapunov-based analysis, we established rigorous conditions under which the closed-loop dynamics achieve FTS and FxTS. In particular, the use of non-Lipschitz and dual-exponent feedback structures enables convergence within finite and uniformly bounded time, respectively.

The efficacy of the discussed methods is illustrated via a nonlinear coupled-tank system, highlighting improved convergence speed and robustness compared to classical approaches.

Overall, the presented framework provides a systematic foundation for fast optimization-driven control and can be extended to more complex and distributed systems.

References

1. Hauswirth, A., Bolognani, S., Hug, G., Dörfler, F.: Timescale separation in autonomous optimization. IEEE Trans. Autom. Control **66**(2), 611–624 (2020)
2. Diana, B., Pandey, S., Kamal, S., Dinh, T.N.: Finite and fixed-time feedback-based continuous-time optimization. Automatica **183**, 112569 (2026)

Chapter 8
Concluding Remarks and Future Work

Abstract This chapter concludes the monograph by synthesizing the key contributions of the work, while also acknowledging its inherent limitations. Furthermore, it outlines promising directions for future research and exploration.

8.1 Conclusions

The primary focus of this monograph is to contribute to the advancement of continuous-time optimization, a field that has received significant interest in recent years due to its broad range of applications, including robot motion coordination, optimal power flow computation, and image processing. The individual chapters of this monograph present unique contributions, summarized as follows:

In the framework developed in Chap. 4, first, we introduced a predefined-time convergent dynamics to solve unconstrained optimization problems. This dynamics has the property that its equilibrium point is the optimal point of the optimization problem, and the optimal point is reached in a pre-specified time. As continuous-time dynamics to solve optimization problems are more amenable to mathematical analysis, we use the Lyapunov stability theory approach to rigorously study the predefined-time convergence of the proposed dynamics. Next, we proposed projected PT-GF to solve equality-constrained optimization problems. To rigorously characterize the impact of perturbations on predefined-time gradient flows, we employ the framework of ISS theory. Theoretical analysis and extensive simulation studies are employed to demonstrate the effectiveness and robustness of the discussed approach.

Chapter 5 extended the proposed approach in Chap. 4 to solve optimization whose objective function and constraints are varying with time. First, we introduced a Newton-like predefined-time stable dynamics to solve TVO problems, and its Lyapunov-based convergence analysis is discussed for both unconstrained as well as constrained case. A robust Newton-like predefined-time stable dynamics is designed for instances when the exact rate of change of the gradient of the objective function is unknown Levenberg-Marquardt-like predefined-time stable dynamics is discussed

S. Kamal, S. Pandey, *A Continuous-Time Perspective on Accelerated Optimization and Control*, Studies in Systems, Decision and Control 693,
https://doi.org/10.1007/978-3-032-31645-5_8

for the cases when the Hessian of the objective function is not invertible, with the assumption based on gradient dominance. The convergence analysis of this dynamics is studied in the sense of ISS, where the input is a perturbation from the true Hessian. Numerical examples for the unconstrained and constrained case of TVO problems are demonstrated using the proposed method. We have also shown that the discretized implementation is conterminous with theoretical results.

Chapter 6 of the monograph focuses on exploring the real-world application of the proposed dynamics in Chap. 5. We illustrate the application of the proposed approach in robot navigation with collision avoidance. We explore a situation where the robot is tasked with tracking a moving target in a priori chosen time. This chapter established a novel vector field-based strategy for predefined-time, collision-free navigation of disk-shaped robots in sphere worlds. We guaranteed convergence to a moving target from almost all initial positions by formulating the navigation task as a TVO problem with inequality constraints. The efficacy of our proposed predefined-time convergent dynamics was rigorously validated through simulations, demonstrating robust performance in complex, obstacle-filled environments.

Chapter 7 presented a feedback-based CTO framework with emphasis on FTC and FxTC. By embedding optimization dynamics within system feedback loops, gradient-based, finite-time, and fixed-time controllers were systematically developed. Lyapunov-based analysis provided rigorous guarantees on convergence behavior, highlighting the role of non-Lipschitz and dual-exponent structures in achieving accelerated and uniformly bounded convergence. The efficacy of the discussed approach was illustrated through a coupled-tank system, illustrating improved performance in terms of convergence speed and robustness. These results establish a strong foundation for fast optimization-driven control of nonlinear dynamical systems.

8.2 Scope for Future Work

While this monograph advances the field of CTO, it is important to acknowledge certain limitations inherent in the proposed framework, which in turn open avenues for further investigation. Accordingly, the following sections highlight promising avenues for future research and development.

- While the current work establishes the ISS of the proposed gradient flows in the presence of perturbations, future work will pursue fundamentally stronger robustness guarantees. We propose to develop novel gradient flow algorithms inspired by the principles of sliding mode control (SMC). The core idea is to deliberately introduce carefully designed switching manifolds or discontinuous elements into the flow dynamics. This strategic modification is anticipated to confer insensitivity to matched perturbations, ensuring that the convergence trajectory reaches and remains precisely at the optimal solution point despite such disturbances. This represents a paradigm shift from merely bounding the error (as per ISS) to actively rejecting its influence for a specific, critical disturbance class.

- Extending the applicability of predefined-time gradient flows to tackle more complex and realistic distributed optimization problems. A critical direction involves handling non-smooth objectives, such as those with L_1-regularization, by integrating proximal operators or subgradient methods into the predefined-time convergence framework, particularly for composite objectives. Concurrently, enhancing robustness against real-world imperfections is paramount; this includes developing flows resilient to communication delays (potentially using Lyapunov-Krasovskii techniques) and stochastic noise affecting both gradients and inter-agent communications. Furthermore, the dynamics of real systems necessitate extensions to time-varying objectives and constraints, requiring adaptive mechanisms combined with predefined-time barrier functions to handle drifting optima and evolving feasible sets effectively.
- Tackling non-convex optimization problems, common in areas like distributed deep learning, demands research into how predefined-time flows can accelerate escape from saddle points and integrate effectively with stochastic gradient descent. Underpinning these advances, a more rigorous and unified theoretical foundation is needed, including the development of specialized Lyapunov theories for networked predefined-time systems and the characterization of robustness margins using ideas like ISS.

MIX
Papier aus verantwortungsvollen Quellen
Paper from responsible sources
FSC® C105338

If you have any concerns about our products,
you can contact us on
ProductSafety@springernature.com

In case Publisher is established outside the EU,
the EU authorized representative is:
Springer Nature Customer Service Center GmbH
Europaplatz 3, 69115 Heidelberg, Germany

Printed by Libri Plureos GmbH
in Hamburg, Germany